WORKSHEETS
FOR CLASSROOM OR LAB PRACTICE

WITH CONTRIBUTIONS FROM

Linda C. Russell
Minneapolis Community and Technical College

PREALGEBRA:
AN INTEGRATED APPROACH

Diana L. Hestwood
Minneapolis Community and Technical College

Margaret L. Lial
American River College

PEARSON
Addison
Wesley

Boston San Francisco New York
London Toronto Sydney Tokyo Singapore Madrid
Mexico City Munich Paris Cape Town Hong Kong Montreal

ISBN-13: 978-0-321-54165-9
ISBN-10: 0-321-54165-0

1 2 3 4 5 6 BB 09 08 07 06

CONTENTS

Chapter 1 INTRODUCTION TO ALGEBRA: INTEGERS

1.1   Place Value

Section 1.1 Objectives
1. Identify whole numbers.
2. Identify the place value of a digit through hundred-trillions.
3. Write a whole number in words or digits.

Key Terms
Answer the following questions about the key terms for Section 1.1.

place value system   **digits**   **whole numbers**

1. Write the correct answer.   1. _____A_____

    a.   The **place value** of 6 in
       67,320 is ten-thousands.

    b.   The **place value** of 4 in
       479,502 is millions.

    c.   The **place value** of 2 in
       4,398,627 is hundreds.

    d.   The **place value** of 9 in
       245,987,465 is thousands.

2. The number system we use has   2. _____D_____

    a.   Nine **digits**.

    b.   Ten **digits**.

    c.   One hundred **digits**.

    d.   An endless number of **digits**.

3. Which answer below shows only **whole**   3. _____B._____
   numbers?

    a.   5, 245, $^{-}78$, 0

    b.   0, 5, 78, 599

    c.   ½, 0, $^{-}90$, forty

    d.   one, 7, 12, ¾

Objective 1   Identify whole numbers.

Circle the whole numbers.

1. $^-15$ ④ 4.16 $^-\dfrac{1}{2}$ $\dfrac{1}{3}$ ⓪ $7\dfrac{1}{8}$ $^-2.71$

2. ⑫ $^-1\dfrac{1}{3}$ $^-6$ 1.3337 6.6666 $^-4.30$ ① ⑪

Objective 2   Identify the place value of a digit through hundred-trillions

Give the place value of the digit 6 in each number.

3.   264   3. ~~⦸⦸⦸⦸⦸~~ tens

4.   85,640   4. hundreds

5.   161,559   5. ten thousands

6.   2,347,886   6. ones

7.   639,111,192   7. hundred mil

8.   2,887,695,142   8. thousands

9.   94,164,372,757   9. ten mil

10. From left to right, name the place value for each 0 in this number: 504,680,804,880,731.

10._____

11. From left to right, name the place value for each 0 in this number: 473,032,880,603,047.

11._____

Objective 3   **Write a whole number in words or digits.**

Write each number in words.

12. 789

12._____

13. 4640

13. *four thousand six hundred fourty*

14. 36,080

14._____

15. 160,180

15. *one hundred sixty thousand one hundred eighty*

16. 2,008,303

16._____

17. 20,508,470

17. *twenty million five hundred eight thousand four hundred seventy*

18. 701,207,330

18._____

19. 9,671,000,637

19. *nine billion six hundred seventy one million six hundred thirty seven*

20. 59,504,806,873

20._____

21. 464,110,054,000

21. *four hundred sixty four trillion one hundred ten million fifty four thousand*

22. 5,509,002,017,800

22. _____

Write each number using digits.

23. Thirty-five thousand, ninety-six

23. 35,096

24. Six hundred ninety thousand, four hundred seventy

24. 690,470

25. Seven million, nine hundred forty-six thousand, two

25. 7,946,002

26. Forty-two million, eighty thousand, four hundred ten

26. 420,080,410

27. Nine hundred million, eleven thousand, five hundred

27. 900,110,500

28. Seventy billion, twelve thousand, eight

28. 700,000,012,000

29. One hundred nine billion, nine hundred sixty-six million

29. 109,266,000,000

30. Seventeen trillion, two hundred ninety-eight million, three thousand, six hundred

30. _____

1.2   Introduction to Signed Numbers

Section 1.2 Objectives
1. Write positive and negative numbers used in everyday situations.
2. Graph signed numbers on a number line.
3. Use the < and > symbols to compare integers.
4. Find the absolute value of integers.

Key Terms *Answer the following questions about the key terms for Section 1.2.*

number line   **integers**   **absolute value**

1. **Whole numbers** differ from **integers** in what way?

    1. __C__

    a. Whole numbers include negative numbers ($\dots ^-3, ^-2, ^-1$) and integers do not.

    b. Whole numbers do not include 0 and integers do.

    c. Whole numbers include 0 and all positive numbers (0,1,2,3…) and integers include negative numbers, 0, and all positive numbers ($\dots ^-3, ^-2, ^-1,0,1,2,3\dots$)

    d. Whole numbers and integers are the same.

2. On a **number line**, zero is

    2. _____

    a. Negative.

    b. Positive.

    c. Both positive and negative.

    d. Neither positive nor negative.

3. _____ is the distance between a number and zero and is always either positive or zero.

    3. __absolute value__

Objective 1 — Write positive and negative numbers used in everyday situations.

Write each negative number with a raised negative sign. Write each positive number in two ways (with and without the raised positive sign).

1. A scuba diver descends to a depth of 14 meters.

   1. $^{-}14$ meters

2. Mercury vaporizes at 357°C.

   2. $357, ^{+}357$

3. A corporation has a shortfall of $1.2 million in its annual revenue.

   3. $^{-}1.2$ million

4. A town receives $2\frac{1}{2}$ inches of rainfall overnight.

   4. $2\frac{1}{2}, ^{+}2\frac{1}{2}$

5. A golfer shoots 2 below par on a round of golf.

   5. $^{-}2$ strokes

6. A gasoline-electric hybrid vehicle gets 65 miles per gallon of gas.

   6. $65, ^{+}65$

7. Improved noise-management technology leads to a 3-decibel reduction in the noise produced by an aircraft.

   7. $^{-}3$ decimals

8. A city has 3696 fewer inhabitants this year than it did last year.

   8. -3696

Objective 2 — Graph signed numbers on a number line.

Graph each set of numbers.

9. $^{-}3, ^{-}4, 3, ^{-}5$

   9.

10. $0, 2, ^{-}1, 5$

    10.

Objective 3   Use the < and > symbols to compare integers.

Write < or > between each pair of numbers to make a true statement.

11. 9 $>$ ⁻6

12. ⁻3 $<$ 8

13. ⁻7 $<$ ⁻5

14. ⁻10 $<$ 2

15. 0 $>$ ⁻2

16. ⁻1 $>$ ⁻2

17. ⁻10 $<$ 4

18. 2 $<$ 5

19. ⁻1 $<$ 5

20. 8 $>$ 2

21. 4 $>$ 0

22. ⁻3 $<$ 2

23. 5 $>$ ⁻8

24. ⁻4 $>$ ⁻10

Objective 4   Find the absolute value of integers.

Find each absolute value.

25. $\left| ⁻3 \right|$

26. $\left| 2 \right|$

27. $\left| ⁻95 \right|$

28. $\left| ⁻11 \right|$

29. $\left| 926 \right|$

30. $\left| ⁻474 \right|$

25. ___3___

26. ___2___

27. ___95___

28. ___11___

29. ___926___

30. ___474___

1.3   Adding Integers

Section 1.3 Objectives
1. Add integers.
2. Identify properties of addition.

Key Terms Answer the following questions about the key terms for Section 1.3.

addends   **sum**   **addition property of 0**

commutative property of addition   **associative property of addition**

1. When you **add** numbers together, the   1.___A___
   answer is called the

   a. sum

   b. quotient

   c. product

   d. opposite

2. The definition of the **commutative**   2.___C___
   property of addition is

   a. Changing the grouping of addends
       does not change the sum.

   b. Changing the order of two factors
       does not change the product.

   c. Changing the order of two
       addends does not change the sum.

   d. Changing the grouping of factors
       does not change the product.

3. The example, 472 + 0 = 472, illustrates   3.___A___
   which property?

   a. **Addition property of 0.**

   b. **Multiplication property of 0.**

   c. **Multiplication property of 1.**

   d. **Distributive property.**

4. The example, $(5 + 8) + 10 = 5 + (8 + 10)$, illustrates which property?

    4._____

   a. the **distributive property.**

   b. the **associative property of addition.**

   c. the **associative property of multiplication.**

   d. the **commutative property of addition.**

| **Objective 1** | Add integers. |

Add by using the number line.

1. $^-1+3$ *2*

    1. ++++++++++→

2. $^-5+3$

      -2

    2. ++++++++++→

Add

3. $^-5+^-9$ **3.** *-14*

4. $9 + 18$ **4.**_____

5. $^-23+^-51$ **5.** *-74*

6. $67 + 83$ **6.**_____

7. $^-4+2$ **7.** *-2*

8. $5 + {}^-7$ 8._____

9. ${}^-26 + 17$ 9.____-9_____

10. $24 + {}^-20$ 10._____

11. ${}^-23 + {}^-9$ 11.____-32_____

12. $8 + {}^-9$ 12._____

13. ${}^-22 + {}^-6$ 13.____-28_____

14. $7 + {}^-22$ 14._____

15. ${}^-1 + 10$ 15._____9_____

16. $1 + {}^-6$ 16._____

17. $6 + {}^-15$ 17.____-9_____

18. ${}^-12 + {}^-1 + {}^-6$ 18._____

19. ${}^-14 + 0 + 17$ 19._____3_____

20. $2 + 2 + {}^-21$ 20._____

21. ${}^-2 + {}^-16 + {}^-4 + {}^-16$ 21.____${}^-38$_____

22. ${}^-10 + {}^-6 + 1 + 1$ 22._____

Write an addition problem for each situation and find the sum.

23. A checking account was overdrawn by $371 before a $500 deposit was made. What was the balance after the deposit? 23.____$\$129$_____

24. While playing a card game, Diane first gained 23 points, then lost 40 points, and finally gained 11 points. What was her final score? 24.____$23 - 40 + 11$_____

Objective 2 **Identify properties of addition.**

Rewrite each sum, using the commutative property of addition, and find the sum both ways.

25. $^-9 + {}^-2$

25. -11

26. $7 + {}^-4$

26. 3

In each addition problem, write parentheses around the two addends that would be easiest to add. Then find the sum.

27. $^-8 + 8 + {}^-15$

27. -15

28. $^-2 + 4 + 6$

28. _____

Find each sum

29. $348 + {}^-281$

29. 168

30. $^-1785 + 670 + {}^-241$

30. _____

1.4   Subtracting Integers

Section 1.4 Objectives
1. Find the opposite of a signed number.
2. Subtract integers.
3. Combine adding and subtracting of integers.

Key Terms *Answer the following questions about the key terms for Section 1.4.*

Opposite

1. Which statement below shows the   1._____B_____
   additive inverse, also called a number's
   opposite?

   a. The opposite of 7 is 7.

   b. The opposite of ⁻15 is 15.

   c. The opposite of 25 is 25^2.

   d. The opposite of 0 is 1.

Objective 1   Find the opposite of a signed number.

Find the opposite (additive inverse) of each number. Show that the sum of the number and its opposite is 0.

1. 8   1. $-8 + 8 = 0$

2. ⁻5   2. $-5 + 5 = 0$

3. ⁻1   3. $-1 + 1 = 0$

4. 4   4. $-4 + 4 = 0$

Objective 2   Subtract integers.

Subtract by changing subtraction to addition.

5. $8 - 7$

5. _____1_____

6. $10 - {}^-3$

6. _____13_____

7. ${}^-13 - 1$

7. ___ -~~12~~ 14 _____

8. ${}^-18 + {}^+20$

8. _____ +2 _____

9. $10 - {}^-10$

9. _____20_____

10. ${}^-9 - 10$

10. ____ -19 _____

11. ${}^-8 - {}^-20$

11. _____12_____

12. ${}^-7 - 12$

12. _____-19_____

13. ${}^-16 - 3$

13. ____ -19 _____

14. $9 - 4$

14. _____5_____

15. ${}^-8 - {}^-19$

15. _____11_____

16. $10 - 0$ 16. _____ 10 _____

17. $^-2 - {}^-11$ 17. _____ 9 _____

18. $10 - 2$ 18. _____ 8 _____

19 $^-10 - {}^-14$ 19. _____ 4 _____

20. $3 + {}^-6$ 20. _____ 9 _____

21. $^-16 - {}^-1$ 21. _____ -15 _____

22. $1 - 5$ 22. _____ -4 _____

23. $2 - {}^-11$ 23. _____ 13 _____

24. $^-4 + {}^-6$ 24. _____ 2 _____

| Objective 3 | **Combine adding and subtracting of integers.** |

Simplify.

25. $^-20 + 14 - 12$ 25. _____ 6 _____

26. $17 - 6 + 7$ 26. _____ 18 _____

27. $^{-}2-18+20$

27. _____0_____

28. $^{-}11+^{-}4-12$

28. _____-27_____

29. $^{-}13+^{-}14-3$

29. _____-30_____

30. $^{-}1+^{-}13-^{-}7$

30. _____-7_____

1.5   Problem Solving: Rounding and Estimating

Section 1.5 Objectives
1. Locate the place to which a number is to be rounded.
2. Round integers.
3. Use front end rounding to estimate answers in addition and subtraction.

Key Terms *Answer the following questions about the key terms for Section 1.5.*

rounding   **estimate**   **front end rounding**

1. When using front end rounding, the rounded   **1.** _____ C _____
    number:

    a. is all zeroes.

    b. always has exactly one zero.

    c. is all zeroes except the first digit.

    d. is all zeroes except the last digit.

2. What is true about **rounding**?   **2.** _____ D _____

    a. Rounding is more accurate than
        using the original number.

    b. Rounding takes more time work
        problems.

    c. A rounded number is always smaller
        than the original number.

    d. A rounded number is easier to work
        with than the original.

3. Another word for **estimate** is   **3.** _____ B _____

    a. Incorrect.

    b. Approximate.

    c. Unknown.

    d. Untrue.

Objective 1   Locate the place to which a number is to be rounded.

Locate and draw a line under the place to which the number is to be rounded.

1. $^-1\textcircled{3}5$ (nearest ten)

2. $4\textcircled{2}817$ (nearest thousand)

3. $\textcircled{5}132,699$ (nearest million)

4. $58,7\textcircled{0}0,030$ (nearest ten-thousand)

Objective 2   Round integers.

Round each number to the indicated place.

5.   133 to the nearest ten

5. _____ 130 _____

6. $^-867$ to the nearest ten

6. _____ $^-870$ _____

7.   6984 to the nearest hundred

7. _____ 7000 _____

8.   32,006 to the nearest hundred

8. _____ 32000 _____

9. $^-37,782$ to the nearest thousand

9. _____ $-38,000$ _____

10. $^-70,215$ to the nearest thousand

10. _____ $-70,000$ _____

11.   908,310 to the nearest ten-thousand

11. _____ 910,300 _____

12.   895,118 to the nearest ten-thousand

12. _____ 900,000 _____

13. 9,570,329 to the nearest million   **13.** _10,000,000_

14. 6,829,055 to the nearest million   **14.** _7,000,000_

| Objective 3 | **Use front end rounding to estimate answers in addition and subtraction.** |

Use front end rounding to round each number.

15. Clyde withdrew $1786 from his bank account.   **15.** _2,000_

16. Carol weighs 114 pounds.   **16.** _100_

Use your estimation skills to pick the most reasonable answer for each addition. Do not solve the problems. Circle your choices.

17. $^-12 + {}^-76$
Estimate: _$^-10$_ + _$^-80$_ = _$^-90$_
Exact: $^-88$ $^-45$ $^-17$

17.
Estimate _-90_
Exact _$^-88$_

18. $72 + {}^-56$
Estimate: _70_ + _-60_ = _10_
Exact: 48 87 16

18.
Estimate _10 70_
Exact _16_

19. $^-123 + 322$
Estimate:
Exact: $^-8050$ 9416 199

19.
Estimate _300_
Exact _199_

20. $^-7810 + 826$

Estimate:

  Exact: $^-6984$ $^-1709$ 4764

20.

Estimate ___-7200___

Exact ___-6984___

21. $8117 + {}^-5160$

Estimate:

  Exact: 2957 8487 $^-2642$

21.

Estimate ___3000___

Exact ___2957___

22. $^-619 + 9985$

Estimate:

  Exact: $^-9609$ 9366 4263

22.

Estimate ___9400___

Exact ___9366___

Change subtractions to addition. Then use your estimation skills to pick the most reasonable answer for each problem. Circle your choices.

23. $83 - {}^-75$

Estimate:

  Exact: 62 158 $^-54$

23.

Estimate ___160___

Exact ___158___

24. $^-61 - 57$

Estimate:

  Exact: 78 $^-118$ $^-20$

24.

Estimate ___-120___

Exact ___-118___

25. 8610 − 1448
    Estimate:
    Exact: 7162   ⁻3039   ⁻342

25.

Estimate   3000

Exact   7162

26. ⁻5172 − 3850
    Estimate:
    Exact: ⁻9022   3589   7060

26.

Estimate   −9000

Exact   −9022

27. ⁻830 − 8237
    Estimate:
    Exact: ⁻4934   ⁻9067   9366

27.

Estimate   −8,8000

Exact   −9067

28. ⁻5994 − ⁻184
    Estimate:
    Exact: 2595   ⁻5810   129

28.

Estimate   3800

Exact   −5810

First use front end rounding to estimate the answer to each application problem. Then find the exact answer.

29. An oil drilling bit was sunk 112 meters into the ground, then raised 43 meters, and finally lowered 22 meters. What was its final depth?

29.

Estimate

Exact

30. A cannon is raised above the horizontal by an angle of 36°. The cannon is then lowered 8°, then lowered another 17°, and finally raised 24°. What is the cannon's final elevation angle?

30.

Estimate

Exact

1.6   Multiplying Integers

Section 1.6 Objectives
1. Use a raised dot or parentheses to express multiplication.
2. Multiply integers.
3. Identify properties of multiplication.
4. Estimate answers to application problems involving multiplication.

Key Terms *Answer the following questions about the key terms for Section 1.6.*

factors   **product**   **multiplication property of 0**

multiplication property of 1   **commutative property of multiplication**

associative property of multiplication   **distributive property**

1. In addition, **addends** are the numbers being added together. In multiplication, _____ are the numbers being multiplied together.

1._____ *C* _____

   a. Factors

   b. Sums

   c. Products

   d. Divisors

2. Multiplying any number by 0 gives a product of

2._____ *C* _____

   a. The same number you started with.

   b. One.

   c. Zero.

   d. Half of the number you started with.

3. The **distributive property** means that we can correctly rewrite 5(4+1) in the following way.

    a. $5 \cdot 4 + 1$

    b. $5 \cdot 4 + 5 \cdot 1$

    c. $5 + 4 \cdot 5 + 1$

    d. $5 \cdot 4 - 5 \cdot 1$

3._____

4. The example, $472 \cdot 1 = 472$, illustrates which property?

    a. The **addition property of 0**.

    b. The **multiplication property of 0**.

    c. The **multiplication property of 1**.

    d. The **distributive property**.

4._____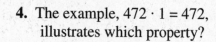C_____

5. The answer to a multiplication problem is called a_____.

5._____

6. Rewriting the example $5(4 \cdot 1)$ as $(5 \cdot 4)1$ is an example of which property?

    e. The **distributive property.**

    f. The **associative property of multiplication.**

    g. The **commutative property of multiplication**.

    h. The **associative property of addition**.

6._____ℓ_____

7. Rewriting the example $5 \cdot 9$ as $9 \cdot 5$ is an example of which property?

7. _____ K _____

   i.   The **distributive property.**

   j.   The **associative property of multiplication.**

   k.   The **commutative property of multiplication.**

   l.   The associative property of addition.

| Objective 1 | Use a raised dot or parentheses to express multiplication. |

Rewrite each multiplication, first using a dot and then using parentheses.

1. $^-3 \times 6$

1. _____ $-3 \cdot 6 \quad -3(6)$ _____

2. $^-7 \times ^-8$

2. _____ $-7 \cdot \, ^-8 \quad -7(-8)$ _____

| Objective 2 | Multiply integers. |

Multiply.

3. $5(^-7)$

3. _____ -35 _____

4. $6(^-8)$

4. _____ $-$ _____

5. $(1)(39)$

5. _____ 39 _____

6. (73)(0)   6. _____0_____

7. (⁻58)(⁻1)   7. ____+58____

8. ⁻86×⁻1   8. ____+86____

9. (⁻6)(⁻5)   9. ____+30____

10. (⁻5)(⁻8)   10. ____+____

11. 27 · ⁻1   11. ____⁻27____

12. ⁻2 · 31   12. ____⁻____

13. ⁻11(⁻2)   13. ____+22____

14. ⁻2(⁻30)   14. ____+60____

15. ⁻2(⁻26)   15. ____+____

16. (⁻48)(1)   16. ____⁻48____

17. ⁻5×⁻1×⁻6   17. ____⁻____

18. $^-7(^-1)(^-4)$

18._____

19. $6(^-9)(3)$

19._____

20. $4(5)(^-5)$

20._____

Fill in each blank to make a true statement.

21. $(^-4)(\underline{\quad}) = ^-24$

21._____

22. $7 \cdot \underline{\quad} = ^-21$

22._____

23. $\underline{\quad} \cdot 9 = ^-36$

23._____

24. $(\underline{\quad})(^-3) = ^-15$

24._____

25. $0 = 18(\underline{\quad})$

25._____

26. $60 = ^-30(\underline{\quad})$

26._____

Objective 3   Identify properties of multiplication.

Rewrite each multiplication, using the stated property. Show that the result is unchanged.

27. (a) Distributive property

$8(4 + ^-7) =$

27.(a)_____

(b) Associative property

$(3 \times ^-2) \times ^-9 =$

(b)_____

28. (a) Commutative property

$^-22 \times 4 =$

(b) Distributive property

$^-5(6 + {}^-1) =$

28.(a)_____

(b)_____

Objective 4   **Estimate answers to application problems involving multiplication.**

First use front end rounding to estimate the answer to each application problem. Then find the exact answer.

29. Jeanne pays someone $8 an hour to work in her yard. If the person works 23 hours over the course of a week, how much does Jeanne owe?

29.

Estimate_____

Exact_____

30. An airplane on an approach path to an airport is losing altitude at the rate of 820 feet per minute. How much altitude does the plane lose in 18 minutes?

30.

Estimate_____

Exact_____

1.7   Dividing Integers

Section 1.7 Objectives
1. Divide integers.
2. Identify properties of division.
3. Combine multiplying and dividing of integers.
4. Interpret remainders in division application problems.

Key Terms *Answer the following questions about the key terms for Section 1.7.*

Quotient

1. The **quotient** is which number in   **1.**_____
   this division: $\frac{14}{2} = 7$?

   a. 14

   b. 2

   c. 7

   d. 1

Objective 1   **Divide integers.**

Divide.

1. $\frac{^-8}{1}$

1. _____-8_____

2. $\frac{21}{^-3}$

2. _____-7_____

3. $\frac{^-9}{^-3}$

3. _____$+3$_____

4. $\dfrac{^-18}{^-3}$

4. _____ +6 _____

5. $\dfrac{28}{^-4}$

5. _____ − 8 _____

6. $\dfrac{12}{6}$

6. _____ + 2 _____

7. $\dfrac{1}{^-1}$

7. _____ − 1 _____

8. $\dfrac{^-16}{2}$

8. _____ − 8 _____

9. $\dfrac{88}{11}$

9. _____ + _____

10. $\dfrac{4}{^-2}$

10. _____ − 2 _____

11. $\dfrac{16}{0}$

11. _____ UD _____

12. $\dfrac{^-24}{0}$

12. _____ − UD _____

13. $\dfrac{^-81}{^-9}$

13. _____ + 9 _____

14. $\dfrac{^-36}{^-12}$

14. _____ + 3 _____

Objective 2   Identify properties of division.

Complete the equation, then state the property represented by the equation.

15. $\dfrac{^-6}{^-6} =$ ___ 1 ___

15. _____ 1 _____

16. $\dfrac{0}{14} =$ ___ Jl ___

16. _____ 11 _____

17. $\dfrac{^-48}{0} =$ ___ UD ___

17. _____ UD _____

18. $\dfrac{33}{1} =$ ___ 33 ___

18. _____ 33 _____

Objective 3   Combine multiplying and dividing of integers.

Simplify.

19. $4 \div {}^-1\left({}^-5\ \right)$

19. _____

20. $0 \div 10(6)$

20. _____ 0/60 _____

21. $32 \div 4\left({}^-99 \div {}^-11 \right)$ **21.** _____

22. $28\left(2 \times {}^-1 \right) \div {}^-8$ **22.** _____

23. ${}^-84 \div {}^-3(3) \div {}^-4$ **23.** _____

24. $1\left({}^-42 \right) \div 7(9)$ **24.** _____

Objective 4   **Estimate answers to application problems involving division**

First estimate the answer to each application problem using front end rounding, then find the exact answer.

25. A company makes 19,200 units of a new children's toy during the first production run. If the total cost of the first run is $364,800, how much did each toy cost?

25.

*Estimate:*_____

Exact: _____

26. A lawn covering 2250 square feet needs fertilizing. The directions say that the fertilizer is to be applied at a rate of 1 pound for every 50 square feet. How many pounds of fertilizer are needed?

26.

*Estimate:*_____

Exact: _____

27. The total cost of a loan, including interest, is $3102. If the loan must be paid off in 6 months, what is the monthly payment?

27.

*Estimate:*_____

Exact: _____

28. There are 12 stages of equal length in a bicycle race. The race covers 168 kilometers in all. How long is each stage?

28.

Estimate:_____

Exact: _____

Objective 5   **Interpret remainders in division application problems.**

Divide, then interpret the remainder in each application.

29. A soda bottling plant produces a run of 1690 soda bottles. Each case of soda holds 20 bottles. How many cases can be filled?

29. 1690/20 _____

30. Computer photo files are being archived onto CDs that each hold 670 megabytes of data. Each photo file contains 5 megabytes and there are 829 files. How many CDs are needed? *Hint: Multiply first*.

30._____

1.8   Exponents and Order of Operations

Section 1.8 Objectives
1. Use exponents to write repeated factors.
2. Simplify expressions containing exponents.
3. Use the order of operations.
4. Simplify expressions with fraction bars.

Key Terms *Answer the following questions about the key terms for Section 1.8.*

Exponent

1. The **exponent** in 5^2 tells us that

1._____  _____

    a.   5 will be used as a factor twice $(5 \cdot 5)$

    b.   5 will be added together $(5 + 5)$

    c.   5 will be added to two $(5 + 2)$

    d.   5 will be multiplied by two $(5 \cdot 2)$

Objective 1   **Use exponents to write repeated factors.**

Rewrite each number in factored form as a number in exponential form, then state how the exponential form is read.

1. (a) $8 \cdot 8 \cdot 8 \cdot 8$

    (b) $9 \cdot 9$

    (c) 10

1.

(a)_____ 8^4 _____

(b)_____ 9^2 _____

(c)_____ 10^1 _____

2. (a) $12 \cdot 12 \cdot 12$

    (b) $6 \cdot 6 \cdot 6 \cdot 6 \cdot 6$

    (c) 5

2.

(a)_____ 12^3 _____

(b)_____ 6^5 _____

(c)_____ 5^1 _____

Objective 2   **Simplify expressions containing exponents.**

Simplify.

3. 3^4

3. $3 \cdot 3 \cdot 3 \cdot 3$ 81

4. 4^3

4. $4 \cdot 4 \cdot 4$ 64

5. $(^-2)^7$

5. $2 \cdot 2 \cdot 2 \cdot 2 \cdot 2 \cdot 2 \cdot 2$ $^-14$

6. $(^-3)^3$

6. $3 \cdot 3 \cdot 3$ ~~27~~

7. $(^-10)^4$

7. $10 \cdot 10 \cdot 10 \cdot 10$ 10,080

8. $(^-5)^4$

8. $5 \cdot 5 \cdot 5 \cdot 5$ 625

9. $2^4 \left(^-3\right)^2$

9. $2 \cdot 2 \cdot 2 \cdot 2 \quad 3 \cdot 3$ 144

10. $\left(^-1\right)^5 \left(^-12\right)$

10. 12

11. $\left(^-11\right)^2 \left(^-1\right)^3$

11. -21

12. $\left(^-3\right)^3 \left(^-2\right)^5$

12. 864

Objective 3 Use the order of operations.

Simplify.

13. $5 + 4\left({}^-3\right)$

13. $\underline{5 + {}^-12 \qquad {}^-27}$

14. $9 - 28 \div 2$

14. $\underline{9-14 \qquad {}^-5}$

15. ${}^-2 + 16 + {}^-8(3)$

15. $\underline{{}^-24 + {}^-2 + 16 \qquad {}^-10}$

16. $10 + {}^-6 + 3\left({}^-1\right)$

16. $\underline{{}^-3 + {}^-6 + 10 \qquad 1}$

17. $3 - {}^-7 + 2^2$

17. $\underline{3 + 7 + 4 \qquad 14}$

18. $5 - {}^-10 + 3^2$

18. $\underline{5 + 10 + 9 \qquad 24}$

19. ${}^-8 + 5(9 - 13)$

19. $\underline{{}^-20 + {}^-8 \qquad {}^-28}$

20. $^-4 + 3(10 - 14)$

20. $-4 + -$

21. $3(^-2 + 6) - (8 - 13)$

21. $^-6 + 18 - 5$

22. $2(3 - 8) - (^-6 + 2)$

22. $6 - 16 + 4$

23. $3 - ^-6\left(^-1\right)^8$

23. $9(-1)^8$

24. $2 + ^+12\left(^-2\right)^2$

24. $14(-2)^2$

25. $^-3(^-5) + 4(8)$

25. $-15 + 4(8)$

26. $5(^-4) + ^-2(^-7)$

26. $-20 + 14$

27. $3\left(4^2\right) + 5(1 + 8) - ^-3$

27.

28. $4\left(5^2\right) - 4(6 + 1) - ^-12$

28.

Objective 4 **Simplify expressions with fraction bars.**

Simplify.

$-2 + 49 + 5$

29. $\dfrac{-2 + 7^2 + -5}{-5 - 11 + 14}$ —

29._____

$-8 + 16 + 9$

30. $\dfrac{-8 + 4^2 - -9}{11 - 3 - 9}$

30._____

31. $\dfrac{-3(4)^2 - 2(5 + -5)}{-3(8 - 13) \div -5}$

31._____

32. $\dfrac{4(2)^2 - 8(7 - 2)}{7(6 - 8) \div 14}$

32._____

Chapter 2 UNDERSTANDING VARIABLES AND SOLVING EQUATIONS

2.1   Introduction to Variables

Section 2.1 Objectives
1. Identify variables, constants, and expressions.
2. Evaluate variable expressions for given replacement values.
3. Write properties of operations using variables.
4. Use exponents with variables.

Key Terms *Answer the following questions about the key terms for Section 2.1.*

> variable   constant   expression
>
> evaluate the expression   coefficient

1. $x + 10$ is an example of   1. _____ D. _____
    a. an **expression**
    b. a **property**
    c. a **constant**
    d. a **variable**

2. "**Evaluate the expression**" means   2. _____ D. _____
    a. decide if it can be solved
    b. give it a rating from 1to5
    c. try to guess what the variable
        amount is
    d. you follow the rule when you
        are given the value of the
        variable

3. A **variable** is written as a letter   3. _____ D _____
    because
    a. It is always the same
    b. It represents a number that can
        change
    c. It is a shorter way to write a
        rule than listing all possible
        numbers
    d. Both b and c of the above

4. A **coefficient** is the number part in an expression that uses only

    a. Division
    b. Addition
    c. Multiplication
    d. Subtraction

4. _____ C _____

5. Which example below is *not* **expression**?

    a. $3x$
    b. $2e$-10
    c. 127 graduates
    d. $11bc^3$

5. _____ C _____

6. A **constant** is

    a. An unknown number
    b. The part of a rule that varies
    c. A number that is added or subtracted in an expression and does not vary
    d. Always 5

6. _____ C _____

Objective 1   Identify variables, constants, and expressions.

Identify the parts of each expression. Choose from these labels: **variable, constant,** *and* **coefficient.**

	variable	constant	coeff.
1. $4z$	Z		4
2. ^-2w	W		-2
3. ^-7+h	h	-7	

var   const.   coeff.

4. *s* - 8

4. $\underline{\quad s \quad\quad -8 \quad\quad}$

5. 3*m* + 11

5. $\underline{\quad m \quad\quad 11 \quad\quad}$

6. ⁻6*n* + 5

6. $\underline{\quad n \quad\quad 5 \quad\quad}$

7. 2*q* + 2

7. $\underline{\quad q \quad\quad 2 \quad\quad}$

8. *pt*

8. $\underline{\quad pt \quad\quad}$

Objective 2   **Evaluate variable expressions for given replacement values.**

Evaluate each expression.

9.   The expression (rule) for the cost in dollars of manufacturing shirts is *n* + 50, where *n* is the number of shirts. Evaluate the expression when
   (a) 70 shirts are manufactured.
   (b) 120 shirts are manufactured.

9.

a. $\underline{ n+50 = 70+50 = 130 \text{ shirts}}$

b. $\underline{ n+50 = 120+50 = 170 \text{ shirts}}$

10.   The expression (rule) for ounces of nitrogen is 3*f*, where *f* is pounds of fertilizer. Evaluate the expression when
   (a) there are 20 pounds of fertilizer.
   (b) there are 64 pounds of fertilizer.

10.

a. $\underline{ 3(20) = }$

b. $\underline{ 3(64) = }$

11. The expression (rule) for the total number of dollars spent on a car, when there is a $2300 down payment and monthly payments of $215, is $2300 + 215t$, where t is the number of monthly payments. Evaluate the expression when
(a) there are 36 monthly payments.
(b) there are 48 monthly payments.

11.
a. _____

b. _____

12. The expression (rule) for the driving hours required for a 975-mile trip by a car is $\frac{975}{v}$, where v is the car's speed in miles per hour. Evaluate the expression when
(a) the speed is 65 miles per hour.
(b) the speed is 75 miles per hour.

12.
a. _____

b. _____

13. The expression (rule) for the amount of insurance paid out on a policy with a $2500 deductible is $\frac{8c}{10} - 250$, where c is total medical costs. Evaluate the expression when
(a) total medical costs are $10,625.
(b) total medical costs are $12,480.

13.
a. _____

b. _____

14. The expression (rule) for the WBGT temperature index is $\frac{7a+3b}{10}$, where a is the temperature measured by a thermometer under a damp cloth and b is the temperature measured by a thermometer inside a black-painted globe. Evaluate the expression when
(a) the wet thermometer measures 58°F and the globe thermometer measures 78°F.
(b) the wet thermometer measures 80°F and the globe thermometer measures 110°F.

14.
a. _____

b. _____

Evaluate each expression for the given replacement values.

15. Evaluate $5x - 2$ when

    (a) x is 3

    (b) x is $^-9$

    (c) x is 1

    (d) x is $^-10$

15.

a. $5(3) - 2 = 13$

b. $5(-9) - 2 = -47$

c. $5(1) - 2 = 3$

d. $5(-10) - 2 = -52$

16. Evaluate $^-4m + 7$ when

    (a) m is $^-1$

    (b) m is 8

    (c) m is $^-6$

    (d) m is 11

16.

a. $-4(-1) + 7 =$

b. $-4(8) + 7 =$

c. $-4(-6) + 7 =$

d. $-4(11) + 7 =$

17. Evaluate $\dfrac{8 - s}{9} + s$ when

    (a) s is $^-1$

    (b) s is $^-10$

    (c) s is 26

    (d) s is 17

17.

a. $\dfrac{8 + {}^+1}{9} = \boxed{1}$

b. $\dfrac{8 + {}^+10}{9} = \boxed{2}$

c. $\dfrac{8 - 26}{9}$

d. $\dfrac{8 - 17}{9}$

18. Evaluate $\dfrac{60}{t} + 2v$ when

    (a) t is 4 and v is $^-10$

    (b) t is $^-5$ and v is $^-4$

    (c) t is 6 and v is 5

    (d) t is $^-12$ and v is 6

18.

a. $\dfrac{60}{4} + 2(-10)$

b. $\dfrac{60}{-5} + 2(-4)$

c. $\dfrac{60}{6} + 2(5)$

d. $\dfrac{60}{-12} + 2(6)$

Objective 3   Write properties of operations using variables.

State the property represented by each equation.

19. $x + 0 = x$

19._____

20. $(1)(k) = k$

20._____

21. $h(a + b) = h(a) + h(b)$

21._____

22. $m \div 0$ is undefined.

22._____

Objective 4   Use exponents with variables.

Rewrite each expression with exponents.

23. $x \cdot x \cdot x$

23._____ x^3 _____

24. $^-6 \cdot y \cdot y$

24._____ $-6y^2$ _____

25. $a \cdot a \cdot b \cdot b \cdot b \cdot b$

25._____ $a^2 b^4$ _____

26. $10(n)(n)(n)(n)$

26._____ $10n^4$ _____

27. $(c)(c)(^-5)(c)$

27._____ $c^2 - 5c$ _____

28. $k \cdot g \cdot {}^-9 \cdot g \cdot k$

28._____ $-9k^2 g^2$ _____

2.2   Simplifying Expressions

Section 2.2 Objectives
1. Combine like terms, using the distributive property.
2. Simplify expressions.
3. Use the distributive property to multiply.

Key Terms: Answer the following questions about the key terms for Section 2.2.

simplifying expressions   **term**

variable term   **like terms**

1. To **simplify** an expression,   1. _____ D _____
    a.   Solve for x
    b.   Add first, then multiply
    c.   Combine all of the numbers
    d.   Combine all the like terms

2. A **term** can   2. _____
    a.   Never have any variables
    b.   Have more than one
         variable
    c.   Never be negative
    d.   Have only one exponent

3. In the following expression, the **like**   3. _____ A _____
    terms are

    $$^-10a^2 + 9a - 2a^2 + 4a^3$$

    a. $^-10a^2$ and $2a^2$
    b. $9a$ and $4a^3$
    c.   there are no like terms
    d.   10, 9, 2 and 4

4. A **variable term** includes   4. _____ A _____
    a.   An exponent and a letter
    b.   A constant and a coefficient
    c.   A coefficient and a letter
    d.   Two letters

47

Name: Date:

Instructor: Section:

Objective 1   Combine like terms, using the distributive property.

Circle the like terms in each expression. Then identify the coefficients of the like terms.

1. $9b + b^2 + b + b^5 + {}^-1$

1. _____

2. $3ac + 2bc + {}^-4ab + ac$

2. _____

Objective 2   Simplify expressions.

Simplify each expression.

3. $^-4v^3 - v^3 + 2v^3$

3. $-3v^2$ _____

4. $5f^4 - 4f^4 - 11f^4$

4. $-9f^4$ _____

5. $24t^3 - 24t^3 + 2t^3$

5. _____

6. $4d^3h + d^3h - 4d^3h$

6. _____

7. $c^3z^4 - 11c^3z^4 - 9c^3z^4$

7. _____

8. $^-11p^3q^3 + 10p^3q^3 + 6p^3q^3$

8. _____

Simplify by combining like terms.

9. $6p + 4p - 11ab$

9. $10p - 11ab$ _____

48

10. $^-12yh + 2dg + 2yh$

10. $\underline{\quad -10yh + 2dg \quad}$

11. $3ht - 5ht + 11$

11. $\underline{\quad -1ht + 11 \quad}$

12. $3dw + 5dw + dw^2$

12. $\underline{\quad 8dw + dw^2 \quad}$

13. $12c^2q - 12cq^2 + 7c - 8$

13. $\underline{\qquad\qquad}$

14. $^-8zt^4 - 8z^2t^2 + 11$

14. $\underline{\qquad\qquad}$

15. $5b^4m + 10 - 2b^4m - 1$

15. $\underline{\qquad\qquad}$

16. $^-50n^4 - 6hq^3 + 11n^4 + 2$

16. $\underline{\qquad\qquad}$

Simplify by using the associative property of multiplication.

17. $^-4(2a)$

17. $\underline{\quad -8a \quad}$

18. $7(^-5x^2)$

18._____

19. $^-5(^-8ab^2)$

19._____

20. $^-6(^-6x^6y^6)$

20.___$36x^6y^4$_____

21. $11(^-zct)$

21.___$-11zct$_____

22. $^-14(^-a^3c^2t)$

22.___$14a^3c^2t$_____

Objective 3   Use the distributive property to multiply.

Use the distributive property to simplify each expression.

23. $2(x+5)$

23.___$2x+10$_____

24. $8(r+2)$

24.___$8r+16$_____

25. $6(z-7)$

25.___$6z-42$_____

26. $5(6a - 7)$

26. $\underline{\quad 30a - 35 \quad}$

27. $^-5(t + 4)$

27. $\underline{\quad -5t - 20 \quad}$

28. $^-2(w + 7)$

28. $\underline{\quad -2w - 14 \quad}$

29. $^-8(p - 11)$

29. $\underline{\quad -8p + 88 \quad}$

30. $^-5(^-2m - 3)$

30. $\underline{\quad +10m + 15 \quad}$

2.3   Solving Equations Using Addition

Section 2.3 Objectives
1. Determine whether a given number is a solution of an equation.
2. Solve equations, using the addition property of equality.
3. Simplify equations before using the addition property of equality.

Key Terms: *Answer the following questions about the key terms for Section 2.3*

equations   solve the equation   solution

addition property of equality   check the solution

1. **Equations** always have
   a. Three parts
   b. The answer, or solution
   c. Two unknown parts
   d. An equal sign

2. The **solution** to an equation replaces the variable and
   a. Must be a coefficient
   b. Never is zero
   c. Makes the equation balance
   d. Cannot be a negative number

3. In the example below, what property is being used when we add 6 to both sides of the equation?

$$b - 6 = {}^-9 + 7$$
$$b - 6 + 6 = {}^-9 + 7 + 6$$
$$b = 4$$

   a. The **addition property of equality**
   b. The **distributive property of equality**
   c. The **simplifying property**
   d. The **commutative property**

4. The best way to **check your solution** is to
   a. Use your calculator
   b. Substitute your solution for the variable and rework it
   c. Use the distributive property
   d. Do the problem twice

1. _____ D _____

2. _____ C _____

3. _____ C _____

4. _____ B _____

Objective 1   **Determine whether a given number is a solution of an equation.**

In each list of numbers, find the one that is a solution of the given equation.

1. $r - 12 = 0$

   $^-8, ^-2, 12$

   1._____12_____

2. $c + 9 = 0$

   $^-9, ^-1, 6$

   2._____-9_____

3. $10 = p + 7$

   $^-7, 3, 7$

   3._____3_____

4. $2 = y - 5$

   $^-2, 2, 4$

   4._____-2_____

Objective 2   **Solve equations using the addition property of equality.**

Solve each equation and check each solution.

5. $q + 6 = 11$

   5._____5_____

6. $b + 2 = 8$

   6._____6_____

7. $4 = s - 9$

7. _____ 13 _____

8. $^-5 = n + 11$

8. _____ 7 _____

Solve each equation.

9. $k + 10 = 1$
$\quad -10 \quad -10$

9. _____ $K = -9$ _____

10. $f + 12 = {}^-4$
$\quad -12 \quad -12$
$\quad\quad\quad -16$

10. _____ $f = -16$ _____

11. $g + 6 = 6$
$\quad -6 \quad -6$
$\quad\quad\quad 0$

11. _____ $g = 0$ _____

12. $b - 7 = {}^-7$
$\quad +7 \quad +7$

12. _____ $b = 0$ _____

13. $^-8 = u - 4$
$\quad +4 \quad +4$

13. _____ $-4 = 4$ _____

14. $^-7 = a + 11$
$\quad {}^-11 \quad {}^-11$

14. _____ $-17 = a$ _____

Check the solution given for each equation. If it is incorrect, find the correct solution.

15. $^-1 + \overset{-1}{v} = {}^-2$

    The solution is $^-1$.

15. _____yes_____

16. $\overset{0}{k} + 10 = 10$

    The solution is 0.

16. _____yes_____

17. $^-6 = \overset{8}{n} - 12$

    The solution is 8.

17. _____6_____

18. $^-10 = 3 + \overset{7}{x}$

    The solution is 7.

18. _____-13_____

Objective 3 Simplify equations before using the addition property of equality.

19. $b - 9 = {}^-6 + 7$

    $b - 9 = 1$
    $\quad +9 \quad +9$

19. _____$b = 10$_____

20. $d - 4 = 11 - 2$

    $d - 4 = 9$
    $\quad +4 \quad +4$

20. _____$d = 13$_____

21. $^-4 + 6 = z + 3$

    $2 = z + 3$
    $-3 \quad\quad -3$

21. _____$z = -1$_____

22. $3 + 11 = h - 5$

    $14 = h - 5$
    $+5 \quad\quad +5$
    19

22. _____$h = 19$_____

23. $8 + p = {}^-6 - 13$

23. _____

24. $1 + u = {}^-4 - 4$

$u = 1$

25. $s + 3 = 2 - 7$

$s + 3 = {}^-5$
$+3 \quad -3$

25. $s = {}^-8$

26. $q - 9 = {}^-12 + 1$

26.

27. $11a - 10a = {}^-5 + 8$

27.

28. ${}^-4v + 6 + 5v = {}^-1 + 9$

$1v + 6 = 8$
$-6 \quad -6$
$\frac{1v}{2} \quad \frac{2}{2}$

28. $v = 2$

29. $8t + 2 - 7t = 1 - 8$

29.

30. ${}^-6r + 5r + 2r - 9 = 10 + 2$

30.

2.4   Solving Equations Using Division

Section 2.4 Objectives
1. Solve equations using the division property of equality.
2. Simplify equations before using the division property of equality.
3. Solve equations such as $^-x = 5$

Key Terms: Answer the following questions about the key terms for Section 2.4.

division property of equality

1. The **division property of equality**   1._____
    a. Allows you to divide by zero
    b. Means that c actually stands for $1c$
    c. Means that you may add the same number to both sides of the equation
    d. Means that an equation will be balanced if you divide each side by the same nonzero number

Objective 1   Solve equations using the division property of equality.

Solve each equation and check each solution.

1. $5s = \dfrac{25}{5}$ 1._____ $s = 5$ _____

2. $^-3m = \dfrac{21}{3}$ 2._____ $s = -7$ _____

3. $^-72 = 8a$
8

3. _____ $a = ^-8$ _____

4. $^-56 = ^-7b$
7

4. _____ $b = ^-8$ _____

Solve each equation.

5. $13g = ^-65$
13

5. _____ $g = ^-5$ _____

6. $^-3r = 72$
$^-3$

6. _____ $r = ^-24$ _____

7. $52 = ^-13x$
13

7. _____

8. $^-48 = 4y$
4

8. _____ $y = ^-12$ _____

9. $^-35 = ^-7v$
7

9. _____ $v = ^-5$ _____

10. $^-180 = ^-12c$
72

10. _____

| **Objective 2** | **Simplify equations before using the division property of equality.** |

Solve each equation.

11. $53 - 8 = {}^-9d$

$$45 = -9d$$
$$\overline{-9}$$

11. $\underline{\quad d = -5 \quad}$

12. ${}^-93 + 9 = 12p$

12. $\underline{\qquad\qquad\qquad}$

13. $5k - 3k = 38$

$$2k = \frac{38}{2}$$

13. $\underline{\quad k = 19 \quad}$

14. ${}^-19h + 4h = {}^-300$

$$+5h = \frac{-300}{-15}$$

14. $\underline{\quad h = -20 \quad}$

15. $4t = {}^-27 - ({}^-27)$

15. $\underline{\qquad\qquad\qquad}$

16. $120 - 120 = 15y$

$$0 = 15y$$

16. $\underline{\quad 0 = 15y \quad}$

17. $^-3n + 9n = 35 - 37 - 40$

$6n = \dfrac{-42}{6}$

17. $n = -7$

18. $3q - 4q = 25 - 13 - 24$

18. _____

19. $^-14f + 10f = 36 + 44$

$-4f = \dfrac{80}{-4}$

19. $f = -20$

20. $^-26g + 11g = ^-90 - 45$

20. _____

21. $27 = ^-3(3g)$

$\dfrac{27}{-9} = \dfrac{-9g}{-9}$

21. $g = -3$

22. $^-32 = 4(8y)$

22. _____

23. $5(^-4k) = 60$

$\dfrac{-20k = 60}{-20}$

23. $k = -3$

24. $^-9(^-2g) = 54$

Objective 3   **Solve equations such as** $^-x = 5$

Solve each equation.

25. $^-10 = ^-f$
     $\frac{}{^-1}\quad\frac{}{^-1}$

25. $\frac{^-10}{^-1}$ _____

26. $4 = ^-v$
     $\frac{}{^-1}\quad\frac{}{^-1}$

26._____

27. $^-g = 5$
     $\frac{}{^-1}\quad\frac{}{^-1}$

27. $\frac{5}{^-1}$ _____

28. $^-17 = ^-w$
     $\frac{}{^-1}\quad\frac{}{^-1}$

28._____

29. $15 = ^-z$
     $\frac{}{^-1}\quad\frac{}{^-1}$

29. $\frac{15}{^-1}$ _____

30. $^{-}b = 19$ **30.**_____

2.5   Solving Equations with Several Steps

Section 2.5 Objectives
1. Solve equations using the addition and division properties of equality.
2. Solve equations using the distributive, addition, and division properties.

Objective 1   **Solve equations using the addition and division properties of equality.**

Solve each equation and check each solution.

1. $^-3g - 18 = {}^-27$

$$+18 \quad +18$$
$$^-3g = {}^-9$$
$$ {}^-3$$

1. _____ $g = 3$ _____

2. $^-3 = {}^-4b - 3$

2. _____

3. $91 = 8w + 19$

$$^-19 \qquad {}^-19$$
$$\frac{72}{8} = 8w$$

3. _____ $w = 9$ _____

4. $^-7a - 11 = 24$

4. _____

Solve each equation.

5. $10z + 4 = {}^-86$

5. _____

6. ${}^-12m - 21 = {}^-105$

$$\begin{array}{r} +21 \quad +21 \\ \hline {}^-10m = {}^-84 \\ \div 2 \end{array}$$

6. $m = 7$

7. ${}^-37 = {}^-4z - 1$

7. _____

8. ${}^-25 = 5s - 5$

8. _____

9. $4m - 19 = {}^-19$

$$\begin{array}{r} +19 \quad +19 \\ \hline 4m = 0 \end{array}$$

9. $4m = 0$

10. $46 = 10g + 16$

10. _____

Solve each equation and check each solution.

11. $2r - 15 = 9r - 43$

$+15+15$

$2r = 9r +\,^{+}28$

$9r = \,^{+}9r$

$11r = \dfrac{28}{11}$

11._____$v = 2.5$_____

12. $^{-}8u + 7 = 5u + 7$

11._____

13. $^{-}15 + 7k = \,^{-}9k + 1$

$+15+15$

$7k = \,^{-}9k + 16$

$+9k\,+9k$

$16k = 16$

11._____$k = 1$_____

14. $^{-}14 + 8d = \,^{-}24 + 3d$

11._____

Solve each equation.

15. $4g + 14 = \,^{-}7g + 47$

$-14-14$

$4g = \,^{-}7g - 33$

$+7g\,+7g$

$11g = \dfrac{-33}{11}$

15._____$g = -3$_____

67

16. $3 + 3p = 7p - 1$

16. _____

17. $^{-}4n + 2 = {}^{-}7n + 8$

$-2-2$

$\overline{{}^{\sim}4n = {}^{-}7n + 6}$

$+7+7$

$\overline{3n = \dfrac{6}{3}}$

17. _____ $n = 2$ _____

18. $14 + 6r = {}^{-}5r + 25$

18. _____

19. $^{-}6d - 26 = {}^{-}8 + 3d$

$+26+26$

$\overline{-6d = 18 + 3d}$

$^{-}3d-3d$

$\overline{9d = \dfrac{18}{9}}$

19. _____ $d = 2$ _____

20. $6z - 19 = 4z - 13$

20. _____

Name: Date:

Instructor: Section:

Objective 2   **Solve equations using the distributive, addition, and division properties.**

Solve each equation and check each solution.

21. $5(h + 3) = {}^-5$

21. $h = -4$

$$5h + 15 = -5$$
$$-15 \quad -15$$
$$\overline{}$$
$$5h = -20$$
$$\overline{5}$$

22. $4(k - 5) = {}^-12$

22. _____

23. $90 = {}^-10(r - 6)$

23. $r = -3$

$$90 = -10r + 60$$
$$-60 -60$$
$$\overline{}$$
$$30 = -10r$$
$$\overline{-10}$$

24. $40 = 5(b + 8)$

24. _____

Solve each equation.

25. $^-12(h-5)=120$

$-12h+60 = \cancel{120}^{11}$

$\underline{\quad -60 \quad -60}$

$\dfrac{-12h}{-12} = \dfrac{60}{-12}$

25. $\underline{\qquad h=-5 \qquad}$

26. $30 = 6(c-1)$

26. $\underline{\qquad\qquad\qquad}$

27. $42 = 3(h+7)$

$42 = 3h + 21$

$\underline{-21 \qquad\quad -21}$

$\dfrac{21}{3} = 3h$

27. $\underline{\qquad h=7 \qquad}$

28. $4(a-1) = 28$

28. $\underline{\qquad\qquad\qquad}$

29. $3(p+6) = {}^-9$

$3p + 18 = -9$

$\underline{\quad -18 \quad -18}$

$3p = \dfrac{9}{3}$

29. $\underline{\qquad p=2 \qquad}$

30. $^{-}30 = ^{-}10(u+5)$

Chapter 3 SOLVING APPLICATION PROBLEMS

3.1   Problem Solving: Perimeter

Section 3.1 Objectives
1. Use the formula for perimeter of a square to find the perimeter or the length of one side.
2. Use the formula for perimeter of a rectangle to find the perimeter, the length, or the width.
3. Find the perimeter of parallelograms, triangles, and irregular shapes.

Key Terms: Answer the following questions about the key terms for Section 3.1.

formula   **perimeter**   **square**   **rectangle**

parallelogram   **triangle**

1. A correct unit for measuring the   1._____D._____
   perimeter of a shape is
   a. lbs
   b. m^2
   c. in^3
   d. ft

2. The characteristics that define a   2._____
   square are
   a. It has 4 sides, the sides are
      of equal length, the angles
      measure 90°
   b. It has 4 sides, opposite sides
      are of equal length, the
      angles measure 90°
   c. It has 4 sides, opposite sides
      are parallel and of equal
      length
   d. It is the same as a rectangle
      only smaller

3. The difference between a **rectangle** 3._____D._____
   and a **parallelogram** is
   a. The **rectangle's** sides are
      not parallel
   b. The angles in a **rectangle**
      must all measure 90°, the
      angles in a **parallelogram**
      may or may not measure 90°

     c. The **parallelogram's** sides
       are all equal, a **rectangle's**
       sides are not equal.
     d. The **parallelogram** has
       more sides than a **rectangle**

4. **Formulas** are rules used to solve
common types of problems.
Formulas are written in a
shorthand form using
     a. Only known numbers
     b. Feet or inches
     c. Square units
     d. Variables

4. _____

5. A formula for finding the
perimeter of a **triangle** with equal-
length sides, can be correctly
written as
     a. $P = s + s + s$
     b. $P = 3s$
     c. $P = s^3$
     d. Both a and b are correctly
      written

5. _____ D _____

| Objective 1 |

Use the formula for perimeter of a square to find the perimeter
or the length of one side.

Find the perimeter of each square, using the appropriate formula.

1.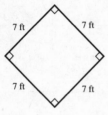
   7 ft   7 ft
   7 ft   7 ft

1. _____ 28 ft. _____

2.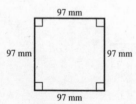
   97 mm
   97 mm   97 mm
   97 mm

2. _____

3. A square book cover measuring
10 inches wide.

3. _____ 40 in. _____

4. A square parking lot measuring 20 **4.**_____
meters on each side.

For the given perimeter of each square, find the length of one side using the appropriate formula.

5. The perimeter is 104 ft.   **5.**_____ 26 ft. _____

6. The perimeter is 44 yd.   **6.**_____

7. A square picture frame with   **7.**_____ 19 in. _____
perimeter 76 in.

8. A square garden with perimeter **8.**_____
64 m.

Objective 2   Use the formula for perimeter of a rectangle to find the perimeter, the length, or the width.

Find the perimeter of each rectangle, using the appropriate formula.

9.   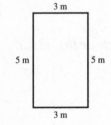   **9.**_____ 16 m _____

10.

13 in.

2 in. [   ] 2 in.

13 in.

10._____

11.

11 yd

6 yd

$$\begin{array}{r} 22 \\ 12 \\ \hline 34 \end{array}$$

11. _34 yd._

12.   8 cm

5 cm

12. _____

13. An 11 in. by 17 in. rectangular piece of
paper

$$\begin{array}{r} 22 \\ 34 \\ \hline 56 \end{array}$$

13. _56 in._

14. A rectangular frozen dinner 20 cm
long by 12 cm wide

14. _____

*For each rectangle, you are given the perimeter and either the length or the width.
Find the unknown measurement by using the appropriate formula.*

15. The perimeter is 40 in. and the width
is 8 in.

8

8

15. _12 in._

16. The perimeter is 90 ft and the length
is 32 ft.

16. _____

17. The length is 14 m and the perimeter
is 50 m.

14   14

17. _11 m_

18. The width is 17 cm and the perimeter is 76 cm.

18._____

19. A 22 ft long rectangular living room has a perimeter of 78 ft.

19._____ 17 ft _____

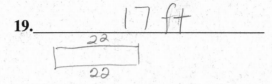

20. A 1 m wide rectangular window has a perimeter of 8 m.

20._____

Objective 3   **Find the perimeter of parallelograms, triangles, and irregular shapes.**

21.

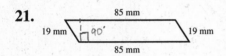

85 mm
19 mm   90°   19 mm
85 mm

21._____ 200 mm _____

22.

18 ft
12 ft   12 ft
18 ft

22._____

23. Parallelogram

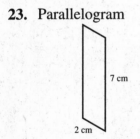

7 cm

2 cm

23._____ 18 cm _____

24. Parallelogram

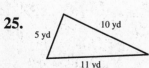

8 in.   3 in.

24._____

25.

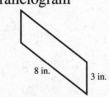

10 yd
5 yd
11 yd

25._____ 26 yd _____

26.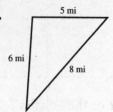

_____

27.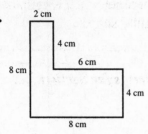

27._____ 32 cm _____

28.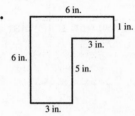

28._____

29.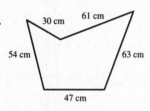

29._____ 285 cm. _____

30.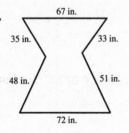

30._____

Name: Date:
Instructor: Section:

3.2   Problem Solving: Area

Section 3.2 Objectives
1. Use the formula for perimeter of a square to find the perimeter or the length of one side.
2. Use the formula for perimeter of a rectangle to find the perimeter, the length, or the width.
3. Find the perimeter of parallelograms, triangles, and irregular shapes.

Key Terms: Answer the following questions about the key terms for Section 3.2.

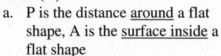
Area

6. The difference between **perimeter (P)**   6._____A_____
    and **area (A)** is
    a.   P is the distance <u>around</u> a flat
         shape, A is the <u>surface inside</u> a
         flat shape
    b.   P is the distance <u>across</u> a flat
         shape, A is the distance <u>around</u> a
         flat shape
    c.   A is the distance <u>around</u> a flat
         shape, P is the <u>surface inside</u> a
         flat shape
    d.   P is the <u>amount</u> of liquid a
         container can hold, A is the
         <u>weight</u> a container can hold

Objective 1   Use the formula for area of a rectangle to find the area, the length, or
                   the width.

Find the area of each rectangle using the appropriate formula.

1.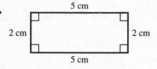

1._____$10 cm^2$_____

2.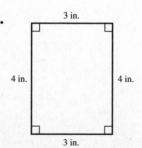

2._____$12 cm^2$_____

3. A rectangular window that measures 3 ft by 7 ft

3. _____ 21 ft.² _____

4. A rectangular table that is 5 m long and 3 m wide

4. _____ 15 m² _____

Use the area of each rectangle and either its length or width, and the appropriate formula, to find the other measurement.

5. The area of a photograph is 660 in.², and the length is 30 in. Find its width.

5. _____ 22 in _____

6. The area of a food tray is 4104 cm² and the width is 54 cm. Find its length.

6. _____ 76 cm _____

7. A rose garden is 40 m wide and has an area of 2400 m². Find its length.

7. _____ 60 m _____

8. A 15 yd² window has a length of 5 yd. Find its width.

8. _____ 3 yd. _____

| **Objective 2** | Use the formula for area of a square to find the area, or the length of one side. |

Find the area of each square using the appropriate formula.

9.

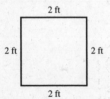

9. _____ 16 ft. _____

10.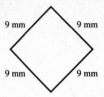

9 mm   9 mm

9 mm   9 mm

10. _____ 81 mm _____

11. A square window pane 33 cm on a side

11. _____ 1089 cm _____

12. A 20 mile square wildlife preserve

12. _____ 400 miles _____

Given the area of each square, find the length of one side by inspection

13. A square cookie sheet has an area of 400 cm^2.

13. _____ 100 cm _____

14. A square garden has an area of 289 m^2.

14. _____ 72.25 m _____

15. The area of a square pot holder is 49 in.^{2}.

15. _____ 12.25 in _____

16. The area of a square game board is 4 ft^2.

16. _____ 1 ft _____

Objective 3	Use the formula for area of a parallelogram to find the area, the base, or the height.

Find the area of each parallelogram using the appropriate formula.

17.

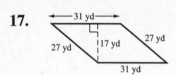

17. $(17)(31) = 527 \text{yd}^2$

18.

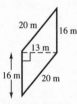

18. $(13)(16) = 208 \text{ m}^2$

19. A parallelogram with height 20 in. and base 23 in.

19. $(20)(23) = 460 \text{ in}^2$

20. A parallelogram measuring 42 miles on the base and 5 miles on the height.

20. $(42)(5) = 210 \text{ m}^2$

Use the area of each parallelogram and either its base or height, and the appropriate formula, to find the other measurement.

21. The area is 204 in.^{2}, and the base is 17 in. Find the height.

21.   12 in.

22. The area is 72 m^2, and the height is 12 m. Find the base.

22.   6 m.

23. The area is 18 yd^2, and the base is 2 yd. Find the height.

23.   9 yd.

24. The area is 460 mm^2, and the height is 46 mm. Find the base

24.   10 mm.

Name: Date:

Instructor: Section:

Objective 4   **Solve application problems involving perimeter and area of rectangles, squares, or parallelograms.**

Find both the perimeter and the area of parallelogram and the rectangle.

25.

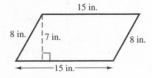

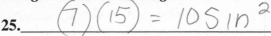

25. $(7)(15) = 105 \text{ in}^2$

26.

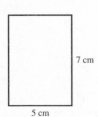

7 cm

5 cm

26. 35 cm^2

Solve each application problem. You may need to find the perimeter, the area, or one of the side measurements.

27. A soybean field is 332 m wide and 512 m long. Find the perimeter and area of the field.

332 m

512 m

27. $a = 169984 \text{ m}^2$

$p = 1,688 \text{ m}$

28. Advertising space on the side of a bus measures 4 ft tall and 22 ft wide. Find the perimeter and area of a poster that would fit there.

22 ft

4 ft

28. $a = 88 \text{ ft}^2$

$p = 52 \text{ ft}$

29. Dennis and Natalie plan to build a patio that has an area of 154 yd^2 along the back edge of their house. If the back edge of their house measures 14 yd long, how far out from their house will the patio extend?

29. _____ 11 yards. _____

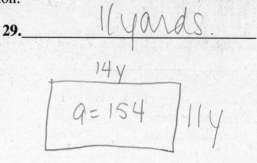

30. Tabitha wants to sew a quilt measuring 88 ft^2 in area. If the quilt must be 11 ft long, how wide will the quilt be?

30. _____ 8 ft _____

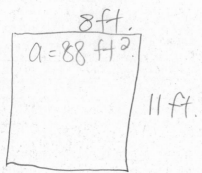

Name: _____ Date: _____

Instructor: _____ Section: _____

3.3 Solving Application Problems with One Unknown Quantity

Section 3.3 Objectives
1. Translate word phrases into algebraic expressions.
2. Translate sentences into equations.
3. Solve application problems with one unknown quantity.

Objective 1 Translate word phrases into algebraic expressions.

Write an algebraic expression using x as the variable.

1. The sum of a number and 17
   1. $X + 17$

2. 57 more than a number
   2. $X + 57$

3. 35 minus a number
   3. $X - 35$

4. A number decreased by 17
   4. $X - 17$

5. Subtract 15 from a number
   5. $15 - X$

6. 3 less than a number
   6. $X - 3$

7. A number subtracted from −2
   7. $-2 - X$

8. 9 fewer than a number
   8. $X - 9$

9. −47 times a number
   9. $X \cdot -47$

10. The product of −6 and a number
    10. $X \cdot -6$

11. 11 fewer than eight times a number   **11.** $8X - 11$

12. 17 less than nine times a number   **12.** $9X - 17$

13. The sum of four times a number and the number   **13.** $4X - X$

14. Twice a number subtracted from the number   **14.** $2X - X$

| Objective 2 | **Translate sentences into equations.**

Translate each sentence into an equation and solve it. Check your solution by going back to the words in the original problem.

15. If three times a number is decreased by eight, the result is 58. Find the number.   **15.** $3X + 8 = 58$

16. The sum of three and seven times a number is 31. Find the number.   **16.** $3 + 7 \cdot X = 31$

17. If seven times a number is subtracted from twelve times the number, the result is –30. Find the number.   **17.** $7X - 12X = -30$

18. If the product of some number and three is increased by 24, the result is five times the number. Find the number.   **18.**
$$x3n + 24 = 5n$$
$$-3n \qquad +3n$$
$$\frac{24}{2} = \frac{2n}{2}$$
$$\boxed{n = 12}$$

19. When seven times a number is subtracted from 61, the result is 13 plus the number. What is the number?

19. $61 - 7X = 13 + X$
$$+ 7X \qquad +7X$$
$$561 = 13 + 8X$$
$$-13 \quad -13$$
$$48 = 8X$$
$X = 6$

20. When nine times a number is decreased by 7, the result is the number increased by 41. Find the number.

20. $9X - 7 = X + 41$

Objective 3   Solve application problems with one unknown quantity.

Translate each sentence into an equation and solve it. Check your solution by going back to the words in the original problem.

21. Melissa gained 12 pounds during a vacation. She then lost 33 pounds before gaining back 6 pounds. If her final weight was 117 pounds, how much did she weigh originally?

21. $12 - 33 + 6 = 117$

22. Shawn was paid a $100 bonus check at work, which he promptly deposited into his checking account. After buying $47 worth of groceries and concert tickets that cost a total of $64, he found that he had a balance of $219 in his account. How much was in his account before he deposited the check?

22. $X + 100 - 47 - 64 = 219$
$$X - 11 = 219$$
$$X = \$230$$

23. There were 12 celery sticks in the vegetable crisper in Lance's refrigerator. After eating some of these celery sticks, Lance added 16 carrot sticks to the crisper. Later, Lance noticed there were a total of 21 celery and carrot sticks altogether. How many celery sticks did Lance eat?

23. $12 - X + 16 = 21$
$$28 - X = 21$$
$$+X \qquad +X$$
$$28 = 21 + X$$
$$-21 \quad 21$$
$$7 = X$$

87

24. Betty left 9 pounds of bird seed in her garage one evening. During the night, mice ate some of the seed. Later, Betty bought a 30-pound bag of bird seed and added it to the seed left in the garage. At this point, she had 36 pounds of seed. How much bird seed did the mice eat?

24.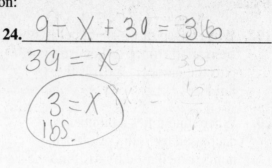

$$9 + X + 30 = 36$$
$$39 = X$$
$$3 = X \text{ lbs.}$$

25. While grocery shopping, Tomba spent $9 less than four times what Pepe spent. If Tomba spent $23, how much did Pepe spend?

25.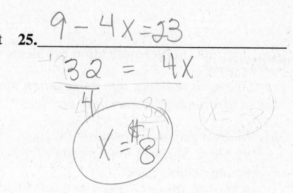

$$9 - 4X = 23$$
$$32 = 4X$$
$$X = \$8$$

26. Margie is 41 years old. She is 15 years less than seven times her daughter's age. How old is Margie's daughter?

26.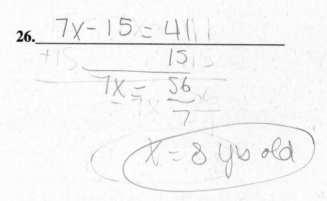

$$7X - 15 = 41$$
$$+15 \qquad 15$$
$$7X = 56$$
$$7$$
$$X = 8 \text{ yrs old}$$

27. Before relatives came to visit, Hilbert bought nine loaves of bread, each of which contained the same number of slices. For breakfast, he served 14 people two pieces of toast each. If Hilbert found that he had 170 slices of bread left for lunch, how many slices of bread were there per loaf?

27.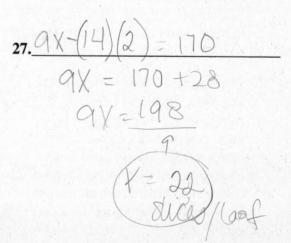

$$9X - (14)(2) = 170$$
$$9X = 170 + 28$$
$$9X = 198$$
$$9$$
$$X = 22 \text{ slices/loaf}$$

28. Before a camping weekend, Angela bought 4 packages of paper plates. On Saturday, one package was used up while on Sunday 30 paper plates were used. If two packages plus 25 plates remain, how many paper plates were there in each package?

28. $4X - X + 30 = 2X + 25$
$3X + 30 = 2X + 25$
$3X - 2X = 30 + 25$
$X = 55$ plates/pkg.

29. The number of bananas an adult gorilla ate is 15 less than three times the number a younger gorilla ate. If the adult gorilla ate 21 bananas, how many did the younger gorilla eat?

29. $15 - 3X = 21$
$\dfrac{36}{3} = \dfrac{3X}{3}$
$X = 12$ bananas

30. The speed that Jacob drives while auto racing is one mile per hour less than twice the speed he drives going home on the freeway. If Jacob races at 139 miles per hour, how fast does he drive home?

30. $2X - 1 = 139$
$2X = \dfrac{140}{2}$
$X = 70 \, mph$

3.4   Solving Application Problems with Two Unknown Quantities

---

Section 3.4 Objectives

1. Solve application problems with two unknown quantities.

Objective 1   Solve application problems with two unknown quantities.

Solve each application problem by using the six problem-solving steps.

1. Debbie is 5 years older than Lori.   1. $5X = 65$
    The sum of their ages is 65. What are
    their ages?

2. The vote totals for the two candidates   2. _516/2_
    for county judge, Jones and Grear,
    differed by only
    516 votes. If there were 16,786 voters
    in the county, how many votes did
    each candidate receive?

3. Last year Lucas earned $5000 less   3. _5000_
    than his wife. Together they earned
    $75,320. How much did each of them
    earn?

4. A $15,000 investment is to be   4. _____
    divided among two stocks so that
    $3000 less is invested in one stock
    than the other. How much will be
    invested in each stock?

5. Linda paid three times as much for   5. _____
    her dog as she did for her cat. If she
    paid a total of $225 for both, how
    much did she pay for each pet?

6. Four times as many customers visited the hardware store on Saturday than on Sunday. If a total of 520 visited during the two days, how many customers visited on each day?

6._____

7. A string is 89 cm long. Bob's cat, Reginald, bit the string into two pieces so that one piece is 17 cm longer than the other. Find the length of both pieces.

7._____

8. Two people are each seated on the two end of a 13-foot wooden plank. The plank balances as a seesaw. If one person is 3 feet further from the balance point than the other, how far is each person from the balance point?

8._____

9. A telephone cable 98 meters in length is cut into two pieces. If one piece is 22 meters longer than the other, how long are the two pieces?

9._____

10. A table top 375 cm wide is cut into two sections so that one section is 49 cm wider than the other section. Find the width of each section.

10._____

11. In an apartment complex, there are 55 fewer females than there are males. If the total number of apartment dwellers is 137, how many males and how many females live in the apartment complex?

11._____

12. The record low temperature in the southern region of a country is 65 degrees higher than the record low in the northern region. If the sum of the two record lows is –105, what is the record low for each region?

12._____

13. An 8-foot long submarine sandwich is to be cut into three pieces. Two pieces are the same length, and the third is two ft longer than each of the other two. Find the length of each piece of the sandwich.

13._____

14. After a 14-meter tall tree is chopped down, it is cut into four pieces. Three pieces are the same length, while the fourth piece is 2 m longer than each of the other three. Find the length of each piece.

14._____

Make a sketch to help you solve each problem. Sketches may vary; show your sketches to your instructor.

15. The perimeter of a rectangular park is 50 miles. If the width is 7 miles, find the length of the park.

15._____

16. The length of a rectangular circuit board is 483 mm, and the perimeter 1590 mm. Find the width of the board.

16._____

17. A rectangular garden is three times as long as it is wide. The perimeter of the garden is 96 yd. Find the length and the width of the garden.

17._____

18. The site of a new building is rectangular in shape. The length is twice the width. If it will require 642 meters of fence to enclose the site, find the length and the width of the new building site.

18._____

19. The length of a picture frame is 5 in. less than three times the width. If the perimeter of the frame is 86 in., find the length and the width of the picture frame.

19._____

20. The perimeter of a rectangular living room floor is 22 yd. If the length of the floor is 3 yd more than the width, find the length and the width of the floor.

20._____

21. A rectangular garden measures 9 m by 7 m. If a 1 m wide walkway that encloses the garden is constructed, what is the area of the walkway and its outside perimeter. Make a sketch.

21._____

22. A rectangular swimming pool is enclosed by a 4-foot wide cement border. If the entire swimming area measures 43 ft by 24 ft, what is the perimeter and the area of the swimming pool alone? Make a sketch.

22._____

Chapter 4 RATIONAL NUMBERS: POSITIVE AND NEGATIVE FRACTIONS

4.1   Introduction to signed fractions

Section 4.1 Objectives
1. Use a fraction to name the part of a whole that is shaded.
2. Identify numerators, denominators, proper fractions, and improper fractions.
3. Graph positive and negative fractions on a number line.
4. Find the absolute value of a fraction.
5. Write equivalent fractions.

Key Terms: Answer the following questions about the key terms for Section 4.1.

fraction   numerator   denominator

proper fraction   improper fraction   equivalent fractions

1. A **fraction**, such as $\dfrac{a}{b}$, always   1. _____C._____
   a. Has a smaller number on the
      top and a larger number on
      the bottom
   b. Represents a number
      smaller than one
   c. Has a nonzero number on
      the bottom
   d. Is a positive number

2. **Proper fractions**   2. _____B._____
   a. Represent a number smaller
      than one
   b. Have a numerator which is
      larger than the denominator
   c. Have a numerator which is
      the same as the denominator
   d. Equal one

3. A **denominator**, such as the 8 in   3._____**C.**_____

   $\frac{3}{8}$,

   a. Can never be one
   b. Is always larger than the
      numerator
   c. Shows that a whole is
      divided into three equal
      parts
   d. Show that a whole is
      divided into eight equal
      parts

4. The **numerator** in a **proper   4._____**B.**_____
   fraction** is
   a. Never one
   b. Smaller than the
      denominator
   c. The same as the
      denominator
   d. Larger than the denominator

5. **Equivalent fractions**   5._____**a.**_____
   a. Represent the same point on
      a number line
   b. Look exactly the same, such
      as $\frac{3}{8}$ and $\frac{3}{8}$
   c. Must be proper fractions
   d. Cannot be proper fractions

Objective 1   Use a fraction to name the part of a whole that is shaded.

Write the fractions for the shaded portion and the unshaded portion of each figure.

1.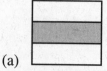
                                       1.

    (a)   (a)_____**1/3**_____

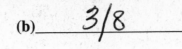

(b)_____ 3/8 _____

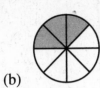

(b)

(c)

(c)_____ 5/6 _____

2.

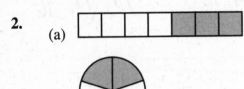

(a)

2.

(a)_____ 3/7 _____

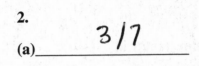

(b)

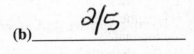

(b)_____ 2/5 _____

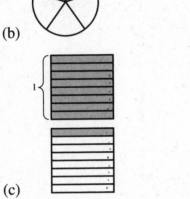

(c)

(c)_____ 9/16 _____

Objective 2 **Identify numerators, denominators, proper fractions, and improper fractions.**

Identify the numerator and denominator in each fraction.

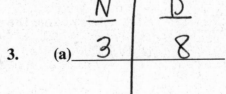

3. (a) $\frac{3}{8}$

3. (a)

$$\frac{N}{3} \quad \frac{D}{8}$$

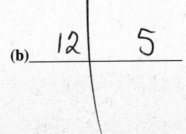

(b) $\frac{12}{5}$

(b) 12 | 5

Name:

Instructor:

Date:

Section:

4. (a) $\dfrac{1}{6}$

(b) $\dfrac{5}{3}$

4.

(a)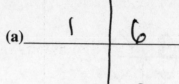

N	D
1	6

(b) S | 3

List the proper and improper fractions in each group of numbers.

Proper   Improper

5. $\dfrac{8}{3}, \dfrac{1}{4}, \dfrac{7}{8}, \dfrac{5}{5}, \dfrac{11}{2}, \dfrac{5}{16}$

8/3, 1/4, 7/8,   5/5, 11/2, 5/16

6. $\dfrac{1}{5}, \dfrac{7}{9}, \dfrac{17}{15}, \dfrac{11}{7}, \dfrac{6}{6}, \dfrac{9}{10}$

1/5, 7/9, 9/10 $\dfrac{17}{15}$ $\dfrac{6}{6}$ $\dfrac{11}{7}$

Objective 3   **Graph positive and negative fractions on a number line.**

Graph each pair of fractions on a number line.

7. $\dfrac{1}{5}, -\dfrac{1}{5}$

7.

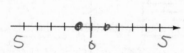

8. $-\dfrac{1}{4}, \dfrac{1}{4}$

8.

9. $-\dfrac{5}{8}, \dfrac{5}{8}$

9.

10. $\dfrac{2}{3}, -\dfrac{2}{3}$

10.

Objective 4   **Find the absolute value of a fraction.**

Find each absolute value.

11. $\left| -\dfrac{3}{5} \right|$ 11. _____3/5_____

12. $|0|$ 12. _____0_____

13. $\left| -\dfrac{15}{7} \right|$ 13. _____15/17_____

14. $\left| \dfrac{8}{9} \right|$ 14. _____8/9_____

15. $\left| -\dfrac{3}{3} \right|$ 15. _____3/3_____

16. $\left| \dfrac{1}{2} \right|$ 16. _____1/2_____

Objective 5   **Write equivalent fractions.**

Rewrite each fraction as an equivalent fraction with a denominator of 48

17. $\dfrac{1}{2}$ 17. _____$\dfrac{24}{48}$_____

18. $\dfrac{1}{3}$ 18. _____

19. $\dfrac{2}{3}$ 19. _____$\dfrac{8}{48}$_____

20. $\dfrac{3}{4}$ 20. _____

21. $\dfrac{1}{6}$ 21. _____$\dfrac{8}{48}$_____

22. $\dfrac{1}{8}$

22._____

23. $\dfrac{3}{8}$

23._____ $\dfrac{2}{48}$ _____

24. $\dfrac{5}{8}$

24._____

Rewrite each fraction as an equivalent fraction with a denominator of 5.

25. $-\dfrac{2}{10}$

25._____ $-1/2$ _____

26. $-\dfrac{4}{20}$

26._____ $-1/5$ _____

27. $-\dfrac{10}{25}$

27._____ $-2/5$ _____

28. $-\dfrac{18}{30}$

28._____ $-$ _____

29. $-\dfrac{12}{15}$

29._____ $-\dfrac{4}{3}$ _____

30. $-\dfrac{14}{35}$

30._____

Name: Date:
Instructor: Section:

4.2   Writing Fractions in Lowest Terms

Section 4.2 Objectives
1. Identify fractions written in lowest terms.
2. Write a fraction in lowest terms using common factors.
3. Write a number as a product of prime factors.
4. Write a fraction in lowest terms, using prime factorization.
5. Write a fraction with variables in lowest terms.

Key Terms: Answer the following questions about the key terms for Section 4.2.

lowest terms   **prime number**

composite number   **prime factorization**

1. A **prime number** has   1._____D._____
    _____factor(s), _____
    a. one, itself
    b. two, 1 and 2
    c. three, 1,2, and 3
    d. **two, 1 and itself**

2. The fraction $\dfrac{10}{15}$ is   2._____A._____

    a. Not in lowest terms
    b. An improper fraction
    c. A prime number when it is
        in lowest terms
    d. Equivalent to $\dfrac{3}{5}$

3. The difference between a **prime**   3._____D._____
   number and a **composite number**
   is
    a. Prime numbers have 2
        factors and composite
        numbers have 3 or more
    b. Prime numbers are odd and
        composite numbers are even
    c. Prime numbers are divisible
        by 1; composite numbers
        are not
    d. None of the above

4. All whole numbers except 0 have factors of 1 and themselves.
  a. True
  b. False

4. _____ True _____

5. **Prime factorization** means finding all factors of a number, and
  a. one of them must be 1
  b. one of them must be 2
  c. all of them must be prime
  d. two of them must be prime

5. _____ B _____

Objective 1   **Identify fractions written in lowest terms.**

Are the following fractions in lowest terms? If not, find a common factor of the numerator and denominator (other than 1).

1. **(a)** $-\dfrac{6}{10}$

1. (a) _____ $-3/5$ _____

  (b) $-\dfrac{7}{11}$

  (b) _____ yes _____

  (c) $\dfrac{20}{25}$

  (c) _____ $-4/5$ _____

2. **(a)** $\dfrac{3}{5}$

2. (a) _____ yes _____

  (b) $\dfrac{12}{16}$

  (b) _____ $3/4$ _____

  (c) $-\dfrac{21}{30}$

  (c) _____ yes _____

Objective 2   **Write a fraction in lowest terms using common factors.**

Write in lowest terms

3   (a) $\dfrac{7}{14}$ 3.   (a) _____ 1/2 _____

     (b) $-\dfrac{24}{36}$ (b) _____ $-\dfrac{4}{6}$ _____

     (c) $\dfrac{33}{88}$ (c) _____ $\dfrac{3}{8}$ _____

4.   (a) $\dfrac{5}{20}$ 4.   (a) _____ 1/4 _____

     (b) $\dfrac{30}{42}$ (b) _____

     (c) $-\dfrac{32}{72}$ (c) _____ $\dfrac{16}{36}=\dfrac{4}{9}$ _____

Objective 3   **Write a number as a product of prime factors.**

Find the prime factorization of each number.

5. 27   5. _____ $3 \cdot 3 \cdot 3$ _____

6. 30   6. _____ $5 \cdot 3 \cdot 2$ _____
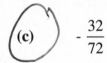

7. 45   7. _____ $3 \cdot 3 \cdot 5$ _____

8. 84   8. _____

9. 105

9. ___5 · 7 · 3___

10. 77

10. ___7 · 11___

11. 104

11. ___2 · 2 · 13___

12. 100
   ∧
  10 10
  ∧   ∧
 5 5 5 5

12. ___5 · 5 · 5 · 5___

Objective 4   **Write a fraction in lowest terms, using prime factorization.**

Write each numerator and denominator as a product of prime factors. Then use the prime factorization to write the fraction in lowest terms.

13. $\frac{12}{42}$

13. $\dfrac{2 \cdot 3 \cdot 3}{7 \cdot 3 \cdot 3}$ $\left(\dfrac{2}{7}\right)$

14. $\frac{9}{45}$

14. $\dfrac{3 \cdot 3}{5 \cdot 3 \cdot 3}$ $\left(\dfrac{1}{5}\right)$

15. $\frac{36}{210}$

15. _____

16. $\frac{18}{99}$

16. $\dfrac{2 \cdot 3 \cdot 3}{3 \cdot 3 \cdot 11}$ $\left(\dfrac{2}{11}\right)$

17. $\frac{63}{105}$

17. _____

18. $\dfrac{16}{180}$

18. _____ $2 \cdot 2 \cdot 2 \cdot 2$ _____

19. $\dfrac{36}{80}$ $\dfrac{4 \cdot 9}{4 \cdot 20}$

19. _____ $\dfrac{3 \cdot 3 \cdot 3 \cdot 3}{2 \cdot 2 \cdot 5 \cdot 5} = \dfrac{9}{20}$ _____

20. $\dfrac{14}{22}$

20. _____ $\dfrac{7 \cdot 2}{7 \cdot 3}$ _____

Objective 5   **Write a fraction with variables in lowest terms.**

Write each fraction in lowest terms.

21. $\dfrac{12x}{28x}$

21. _____ $\dfrac{2x}{4x}$ _____

22. $\dfrac{18y}{24y}$

22. _____ $\dfrac{3}{4}$ _____

23. $\dfrac{15r^2}{18rs}$

23. _____ $\dfrac{3r}{9s}$ _____

24. $\dfrac{6a^2b}{10ab^2}$

24. _____ $\dfrac{3a}{5b}$ _____

25. $\dfrac{5xz^2}{9xyz^2}$

25. $\quad 5/9y$

26. $\dfrac{18cd^2}{18c^2d}$

26. $\quad cd$

27. $\dfrac{25m^2}{16n^2}$

27. $\quad \dfrac{25m^2}{16n^2}$

28. $\dfrac{15rst}{24st}$

28. $\quad \dfrac{5 \cdot 15r}{3 \cdot 24} = \dfrac{5r}{8}$

29. $\dfrac{6xy^2}{35z^3}$

29. $\quad \dfrac{6xy^2}{35z^3}$

30. $\dfrac{16b^2cd}{40b^2d}$

30. $\quad \dfrac{16c}{40}$

4.3 Multiplying and Dividing Signed Fractions

Section 4.3 Objectives
1. Multiply signed fractions.
2. Multiply fractions that involve variables.
3. Divide signed fractions.
4. Divide fractions that involve variables.
5. Solve application problems involving multiplying and dividing fractions.

Key Terms: *Answer the following questions about the key terms for Section 4.3.*

reciprocal

1. The product of the **reciprocal** of a   1._____ D. _____
   negative fraction is
   a. $^-1$
   b. 1
   c. 0
   d. undefined

Objective 1 **Multiply signed fractions.**

Multiply. Write the products in lowest terms.

1. $-\dfrac{5}{8} \times \dfrac{1}{2}$

1. $\dfrac{5 \cdot 1}{2 \cdot 2 \cdot 2 \cdot 2}$

2. $\left(\dfrac{2}{3}\right)\left(-\dfrac{7}{9}\right)$

2. $\dfrac{2 \cdot 7}{3 \cdot 3 \cdot 3}$

3. $\left(-\dfrac{3}{4}\right)\left(-\dfrac{12}{7}\right)$

3. $\dfrac{3 \cdot 3 \cdot 2 \cdot 2}{2 \cdot 2 \cdot 7}$

4. $\dfrac{2}{9} \times \dfrac{12}{5}$

4. $\dfrac{2 \cdot 2 \cdot 2 \cdot 3}{3 \cdot 3 \cdot 5}$

5. $\dfrac{4}{5}$ of 125

5. $\dfrac{4}{5} \cdot \dfrac{125}{1} = \dfrac{2 \cdot 2 \cdot}{5 \cdot 1}$

6. $\dfrac{5}{6}$ of 30

6. $\dfrac{5}{6} \cdot \dfrac{30}{1} = \dfrac{5 \cdot 5 \cdot 3 \cdot 2}{3 \cdot 2 \cdot 1}$

7. $14\left(-\dfrac{2}{7}\right)$

7. $\dfrac{14}{1} \cdot \dfrac{-2}{7} = \dfrac{2 \cdot 7 \cdot 2}{1 \cdot 7}$

8. $-\dfrac{3}{8}(^-24)$

8. $\dfrac{-3}{8}\left(-\dfrac{24}{1}\right) = \dfrac{-3 \cdot 2 \cdot 2 \cdot 2 \cdot 2 \cdot 2}{2 \cdot 2 \cdot 1}$

Objective 2   **Multiply fractions that involve variables.**

Use prime factorization to find these products

9. $\dfrac{3c}{5} \times \dfrac{c}{9}$

9. $\dfrac{3c^2}{45}$

10. $\left(\dfrac{m}{10}\right)\left(\dfrac{25}{m^2}\right)$

10. _____

11. $\dfrac{2x}{5} \times \dfrac{3}{4x}$

11. $\dfrac{6x}{20x}$

12. $\left(\dfrac{2y^2}{6x}\right)\left(\dfrac{3x^2}{y}\right)$

12. _____

13. $\left(\dfrac{2x}{15}\right)\left(\dfrac{3}{4x^2}\right)$

13. $\dfrac{2x \cdot 3}{15 \quad 4x^2}$

14. $\dfrac{5x}{4} \times \dfrac{7}{xy}$

14. _____

Objective 3   **Divide signed fractions.**

Divide. Write the quotients in lowest terms.

15. $\dfrac{3}{7} \div 9$ $\dfrac{3}{7} \cdot \dfrac{1}{9}$

15. $\dfrac{3}{7} \cdot \dfrac{1}{9}$

16. $-\dfrac{7}{8} \div \dfrac{21}{5}$

16. $\dfrac{-7}{8} \cdot \dfrac{5}{21}$

17. $-\dfrac{5}{12} \div \dfrac{15}{8}$

17. $\dfrac{-5}{12} \cdot \dfrac{8}{15}$

18. $-6 \div \left(-\dfrac{1}{2}\right)$

18. $\dfrac{-6}{1} \cdot \dfrac{-2}{1}$

19. $8 \div \left(-\dfrac{2}{5}\right)$

19. $\dfrac{8}{1} \cdot \dfrac{-5}{2}$

20. $\dfrac{7}{8} \div (^-21)$

20. $\dfrac{7}{8} \cdot \dfrac{-21}{1}$

Objective 4   **Divide fractions that involve variables.**

21. $\dfrac{3b}{5a} \div \dfrac{6}{7ab}$

21._____

22. $\dfrac{x}{3y} \div \dfrac{4x}{7y}$

22._____

23. $\dfrac{a^2 b}{c} \div \dfrac{ab^2}{c}$

23._____

24. $\dfrac{3mn}{7} \div \dfrac{6n}{m}$

24._____

25. $10x^2 \div \dfrac{5x}{3}$

25._____

26. $\dfrac{7c}{3d} \div 14c^2 d$

26._____

Objective 5   **Solve application problems involving multiplying and dividing fractions.**

Solve each application problem.

27. How many times can a $1\frac{2}{3}$-quart spray bottle be filled before 20 quarts of water are used up?

27. $\qquad 20 \div 1\frac{2}{3} \qquad$

28. Signe gives $\frac{1}{12}$ of her income to charities. Last year she earned $48,000. How much did she give to charities?

28. $\qquad 48{,}000 \div \frac{1}{2} \qquad$

29. At the Summer Volleyball Tournament, $\frac{5}{8}$ of the players are women. If there are 360 players, how many are women? How many are men?

29. _____

30. Rochelle sells hats at craft shows. She needs $\frac{2}{5}$ yd for each hat. How many hats can she make from 10 yd of fabric?

30. _____

4.4   Adding and Subtracting Signed Fractions

Section 4.4 Objectives
1. Add and subtract like fractions.
2. Find the lowest common denominator for unlike fractions.
3. Add and subtract unlike fractions.
4. Add and subtract unlike fractions that contain variables

Key Terms: Answer the following questions about the key terms for Section 4.4.

like fractions   **unlike fractions**   **least common denominator**

1. Fractions may be added or
   subtracted only when
      a. They have no negative
        numbers
      b. They are in lowest terms
      c. They have the same
        numerators
      d. They have the same
        denominators

   1. _____ D _____

2. $\dfrac{3}{9}$ and $\dfrac{9}{8}$ are **unlike fractions**

   because

      a. $\dfrac{3}{9}$ is not in lowest terms

      b. $\dfrac{9}{8}$ is an improper fraction

      c. they have different
        denominators
      d. they have different
        numerators

   2. _____ C _____

3. An **LCD** must be
      a. The lowest number that can
        be divided evenly by both
        denominators
      b. Prime
      c. The lowest number that can
        be divided evenly by both
        numerators
      d. Either **a** or **c** above

   3. _____ A _____

Objective 1   Add and subtract like fractions.

Write each sum or difference in lowest terms.

1. $\dfrac{1}{7} + \dfrac{6}{7}$

1. $\dfrac{7}{7} = \boxed{1}$

2. $\dfrac{1}{10} + \dfrac{3}{10}$

2. $\dfrac{4}{10} = \boxed{\dfrac{2}{5}}$

3. $-\dfrac{2}{7} - \dfrac{3}{7}$

3. $\dfrac{2}{7}$

4. $-\dfrac{14}{15} + \dfrac{4}{15}$

4. $\dfrac{-10}{15} = \boxed{-\dfrac{3}{5}}$

5. $\dfrac{7}{y^2} - \dfrac{3}{y^2}$

5. $\dfrac{4}{y^2}$

6. $-\dfrac{8}{ab} - \dfrac{2}{ab}$

6. $\dfrac{-6}{ab}$

Objective 2   Find the lowest common denominator for unlike fractions.

Find the LCD for each pair of fractions

7. $\dfrac{3}{7}$ and $\dfrac{3}{14}$

7. $\boxed{14}$

8. $\dfrac{1}{3}$ and $\dfrac{2}{7}$

8. $\boxed{21}$

9. $\dfrac{1}{4}$ and $\dfrac{5}{6}$

9. ⟨12⟩

10. $\dfrac{5}{6}$ and $\dfrac{7}{15}$

10. (30)

11. $\dfrac{4}{15}$ and $\dfrac{8}{21}$

11. $\dfrac{15 - 5 \quad 3}{21 - \quad 3 \quad 7}$ (105)

$5 \cdot 3 \cdot 7$

12. $\dfrac{7}{12}$ and $\dfrac{3}{40}$

12. $12 - 2 \cdot 2 \cdot 3$

$40 -$ ~~2⋅2⋅2~~ $2 \cdot 5 \cdot 2 \cdot 2$

Objective 3   Add and subtract unlike fractions.

Find each sum or difference. Write all answers in lowest terms

13. $\dfrac{2(3)}{10(3)} + \dfrac{2(2)}{15(2)}$

13. $\overset{9}{14} \over 30$ $+ \dfrac{4}{30} = \dfrac{18}{30}$

14. $\dfrac{7}{16} + \dfrac{9}{24}$

14. _____

15. $\dfrac{5(3)}{8(3)} - \dfrac{1(4)}{6(4)}$

15. $\dfrac{15}{24} - \dfrac{4}{24} = \dfrac{11}{24}$

16. $\frac{2}{3} + \frac{1}{4}$

16. $\dfrac{8}{12} + \dfrac{3}{12} = \dfrac{11}{12}$

17. $\frac{3}{10} + \frac{3}{8}$

17. 40

18. $\frac{7}{9} - \frac{1}{6}$

18. 18

19. $\frac{11}{12} - \frac{7}{10}$

19. 60

20. $\frac{7}{12} + \frac{1}{10}$

20. $\dfrac{8}{60}$

21. $\frac{13}{24} - \frac{5}{16}$

21. 48

22. $\frac{7}{15} + \frac{2}{35}$

22. 105

Name: Date:
Instructor: Section:

Objective 4 Add and subtract unlike fractions that contain variables.

Find each sum or difference.

23. $\dfrac{8}{ab} - \dfrac{3}{ab}$

23. $\dfrac{5}{ab}$

24. $\dfrac{7}{8} - \dfrac{3}{y}$

24. $\dfrac{4}{8y} = \dfrac{1}{2y}$

25. $\dfrac{3}{n} - \dfrac{3}{5}$

25. _____

26. $6 - \dfrac{x}{2}$

26. $12x$

27. $-\dfrac{2}{z^2} - \dfrac{y}{z}$

27. $\dfrac{-2y}{z}$

28. $\dfrac{3}{a} - \dfrac{b}{4}$

28. 36

29. $0 - \dfrac{ab}{6c}$

29. $\dfrac{-ab}{6b}$

30. $\dfrac{m}{6} + \dfrac{3}{n}$

30. $3m$

Name:
Instructor:

Date:
Section:

4.5 Problem Solving: Mixed Numbers and Estimating

Section 4.5 Objectives
1. Identify mixed numbers and graph them on a number line
2. Rewrite mixed numbers as improper fractions, or the reverse.
3. Estimate the answer and multiply or divide mixed numbers.
4. Estimate the answer and add or subtract mixed numbers.
5. Solve application problems containing mixed numbers

Key Terms: *Answer the following questions about the key terms for Section 4.5.*

mixed number

1.A **mixed number** may also be written as   1._____ a _____
   a. A **proper fraction**
   b. An **improper fraction**
   c. Two **whole numbers**
   d. A numeral **1** and a **fraction**

Objective 1   Identify mixed numbers and graph them on a number line.

Graph the mixed numbers or improper fractions on a number line.

1. $2\frac{2}{3}$ and $-2\frac{2}{3}$

1.
   -2 -1 0 1 2

2. $1\frac{1}{4}$ and $-1\frac{1}{4}$

2.

3. $\frac{5}{2}$ and $-\frac{5}{2}$

3.
   -3 -2 -1 0 1 2 3

4. $\frac{13}{3}$ and $-\frac{13}{3}$

4.

Objective 2   Rewrite mixed numbers as improper fractions, or the reverse.

Write each mixed number as an improper fraction

5. $2\frac{3}{5}$

5. _____ 13/5 _____

6. $5\frac{3}{7}$

6. _____ 38/7 _____

7. $^-6\frac{3}{8}$

7. _____ -51/8 _____

8. $-4\frac{1}{6}$

8. _____ -25/6 _____

Write each improper fraction as a mixed number in simplest form.

9. $\frac{33}{7}$

9. _____ 4 5/7 _____

10. $-\frac{9}{7}$

10. _____ -1 2/7 _____

11. $-\frac{42}{8}$

11. _____ -5$\frac{1}{4}$ _____

12. $\frac{14}{6}$

12. _____ 2$\frac{1}{3}$ _____

Objective 3   **Estimate the answer and multiply or divide mixed numbers.**

First, round the mixed numbers to the nearest whole number and estimate each answer. Then find the exact answer. Write exact answers in simplest form.

13. $3\frac{2}{3} \times 1\frac{1}{7}$

Estimate:

Exact:

14. $2\frac{1}{2} \times 2\frac{1}{3}$

Estimate:

Exact:

15. $3\frac{3}{4} \div 1\frac{3}{8}$

Estimate:

Exact:

16. $1\frac{1}{3} \div 1\frac{1}{7}$

Estimate:

Exact:

17. $\left(3\frac{1}{8}\right)\left(5\frac{1}{3}\right)$

Estimate:

Exact:

18. $3\frac{1}{2} \div 2\frac{2}{5}$

Estimate:

Exact:

13.

Estimate: $4 \cdot 1 = \boxed{4}$

Exact: $4\ 4/21$

14.

Estimate: $2 \cdot 2 = \boxed{4}$

Exact:

15.

Estimate: $4 \div 1 = \boxed{4}$

Exact: $2\ 8/11$

16.

Estimate: $1 \div 1 = \boxed{1}$

Exact:

17.

Estimate: $3 \cdot 5 = \boxed{15}$

Exact: $16\ 2/3$

18.

Estimate: $3 \div 2$

Exact:

Objective 4   Estimate the answer and add or subtract mixed numbers.

First, round the mixed numbers to the nearest whole number and estimate each answer.   Then find the exact answer. Write exact answers in simplest form.

19. $4\frac{3}{8} - 2\frac{5}{8}$

Estimate:

Exact:

19.

Estimate: $4 - 3 = ①$

Exact: 2

20. $4\frac{7}{9} + 6\frac{5}{9}$

Estimate:

Exact:

20.

Estimate: $5 + 7 = ⑫$

Exact:

21. $2\frac{3}{4} + 3\frac{5}{6}$

Estimate:

Exact:

21.

Estimate: $3 + 4 = ⑦$

Exact: $5\frac{19}{12}$

22. $9 - 5\frac{4}{9}$

Estimate:

Exact:

22.

Estimate: $9 - 5 = ④$

Exact:

23. $5 - 2\frac{1}{4}$

Estimate:

Exact:

23.

Estimate: $5 - 2 = ③$

Exact:

24. $7\frac{5}{12} - 3\frac{11}{18}$

Estimate:

Exact:

24.

Estimate: $7 - 4 = ③$

Exact:

Objective 5 — Solve application problems containing mixed numbers.

First, estimate the answer to each application problem. Then find the exact answer.

25. George's daughter grew $1\frac{1}{3}$ inches last year and $2\frac{1}{5}$ inches this year. How much has her height increased over the two years?

25.

Estimate: $1\frac{1}{3} + 2\frac{1}{5} = 1 + 2 = \boxed{3}$

Exact: _____

26. Katlyn used $2\frac{3}{8}$ packages of nuts in her salad recipe. Each package has $3\frac{1}{6}$ ounces of nuts. How many ounces of nuts did she use in the recipe?

26.

Estimate: $2\frac{3}{8} \div 3\frac{1}{6} = 2 \div 3$

Exact: _____

27. Richard worked 7 hours on Saturday and $5\frac{4}{9}$ hours on Sunday. How much longer did he work on Saturday than on Sunday?

27.

Estimate: $7 - 5\frac{4}{9} = 7 - 5 = \boxed{2}$

Exact: _____

28. A carpenter needs two pieces of lumber measuring $3\frac{1}{4}$ feet and $2\frac{7}{16}$ feet. What is the total length of lumber needed?

28.

Estimate: $3\frac{1}{4} + 2\frac{7}{16} = 3 + 2 = \boxed{5}$

Exact: _____

29. Suppose that a pair of pants requires $3\frac{1}{8}$ yd of material. How much material would be needed for 6 pairs of pants?

29.

Estimate: $3\frac{1}{8} \cdot 6 = 3 \cdot 6 = 18\,\text{yd.}$

Exact: _____

Name: Date:
Instructor: Section: ✓

4.6   Exponents, Order of Operations, and Complex Fractions

Section 4.6 Objectives
1. Use fractions with exponents.
2. Use the order of operations with fractions.
3. Simplify complex fractions.

Key Terms: *Answer the following questions about the key terms for Section 4.6.*

complex fraction

1. $\dfrac{\frac{1}{5}\left(\frac{4}{5}\right)}{3}$ is an example of

1.  *D*

    a. an **LCD**
    b. a fraction in **lowest terms**
    c. an **improper fraction**
    d. a **complex fraction**

| **Objective 1** | Use fractions with exponents. |

Simplify.

1. $\left(-\dfrac{2}{5}\right)^2$

1. *4/25*

2. $\left(-\dfrac{1}{4}\right)^3$

2. *-1/64*

3. $\left(-\dfrac{2}{3}\right)^4$

3. *16/81*

4. $\left(-\dfrac{3}{5}\right)^3$

4. *-27/125*

5. $\left(\dfrac{7}{8}\right)^2$

5. _____ 49/64 _____

6. $\left(-\dfrac{7}{5}\right)^2$

6. _____ 1 24/25 _____

7. $-7\left(\dfrac{2}{5}\right)^2$

7. _____ -1 5/25 _____

8. $\left(\dfrac{1}{3}\right)^3\left(\dfrac{1}{4}\right)^2$

8. _____ 1/432 _____

9. $\left(-\dfrac{2}{3}\right)^2\left(\dfrac{1}{2}\right)^3$

9. _____ 1/18 _____

10. $\dfrac{15}{16}\left(\dfrac{4}{5}\right)^2$

10. _____ 3/5 _____

11. $\left(-\dfrac{2}{3}\right)^3\left(\dfrac{3}{2}\right)^2$

11. _____ -2/3 _____

12. $3\left(-\dfrac{1}{3}\right)^3$

12. _____ -1/9 _____

Objective 2   Use the order of operations with fractions.

Simplify.

13. $\left(\dfrac{3}{4}\right)\left(\dfrac{8}{9}\right)-\dfrac{1}{3}$ 13. _____ 1/3 _____

14. $\left(\dfrac{5}{12}\div\dfrac{1}{2}\right)+\left(\dfrac{3}{7}\cdot\dfrac{7}{9}\right)$ 14. _____ 1'/6 _____

15. $\left(4\dfrac{1}{2}\div\dfrac{3}{4}\right)-\left(5\dfrac{1}{4}\cdot1\dfrac{1}{7}\right)$ 15. _____

16. $\dfrac{7}{8}\div\dfrac{3}{4}-\dfrac{5}{6}$ 16. _____ 3/6 _____

17. $10-5\left(7\dfrac{1}{5}\div2\dfrac{1}{4}\right)$ 17. _____

18. $\dfrac{15}{16}-\dfrac{1}{8}-\left(\dfrac{3}{4}\right)^{2}$ 18. _____ 4/16 _____

19. $\left(\dfrac{7}{8}\cdot\dfrac{4}{9}\right)\div\left(\dfrac{1}{3}+\dfrac{1}{6}\right)$ 19. _____ 7/9 _____

20. $\left(\dfrac{3}{2}\right)^{2}-\dfrac{3}{10}\div\dfrac{1}{5}$ 20. _____

21. $\dfrac{1}{6} + 3\left(5\dfrac{5}{9} - 2\dfrac{1}{3}\right)$

21._____

22. $\left(\dfrac{1}{3}\right)^{2} - \left(-\dfrac{1}{2}\right)^{3}$

22._____

Objective 3   **Simplify complex fractions.**

Simplify

23. $\dfrac{-\frac{5}{8}}{-\frac{5}{16}}$

23._____ 2 _____

24. $\dfrac{\frac{16}{35}}{-\frac{4}{70}}$

24._____ $\dfrac{-280}{35}$ _____

25. $\dfrac{-20}{\frac{2}{5}}$

25._____ -50 _____

26. $\dfrac{-8}{\frac{3}{2}}$

26._____ $\dfrac{-16}{3}$ _____

27. $\dfrac{\frac{8}{9}}{2}$

27._____ $4/9$ _____

28. $\dfrac{-\frac{9}{11}}{11}$

28._____ $-9/121$ _____

29. $\dfrac{\left(\frac{2}{3}\right)^2}{\left(-\frac{4}{5}\right)^2}$

29. _____ 25/36 _____

30. $\dfrac{\left(-\frac{1}{2}\right)^3}{\left(\frac{3}{4}\right)^2}$

$-2/9$

30. _____

4.7   Problem Solving: Equations Containing Fractions

Section 4.7 Objectives
1. Use the multiplication property of equality to solve equations containing fractions.
2. Use both the addition and multiplication properties of equality to solve equations containing fractions.
3. Solve application problems using equations containing fractions.

Key Terms: *Answer the following questions about the key terms for Section 4.7*

multiplication property of equality   **division property of equality**.

1. The **multiplication property of equality** means that

$$\frac{1}{3}c = 7$$

$$\frac{3}{1}\left(\frac{1}{3}\right)c = 7\left(\frac{3}{1}\right)$$

will result in a

   a. Balanced equation
   b. Complex fraction
   c. A mixed number
   d. Complex division problem

1._____ A_____

2. When using the **division property of equality**, the only number that may not be used to divide both sides of the equation by is
   a. A nonzero number
   b. A prime number
   c. Zero
   d. A fraction

2._____ D_____

Objective 1   Use the multiplication property of equality to solve equations containing fractions.

Solve each equation and check each solution.

1. $\frac{1}{6}m = 9$

 1. $\frac{\cancel{6}}{\cancel{1}}m = 9 \cdot 6 = \boxed{54}$

2. $\dfrac{2}{3}a = -6$

2. $\dfrac{3}{2}a = -6$

3. $\dfrac{5}{12} = -\dfrac{4}{3}x$

3. $\dfrac{12}{5} = \dfrac{4}{3}x$

4. $8 = \dfrac{1}{4}y$

4. $8 = \dfrac{4}{1}y$

5. $-30 = \dfrac{5}{6}b$

5. $-30 = \dfrac{6}{5}b$

6. $-\dfrac{9}{2}c = -36$

6. $-\dfrac{2}{9}c = -36$

7. $\dfrac{10}{12} = \dfrac{5}{4}m$

7. $\dfrac{10}{12} = \dfrac{4}{5}m$

8. $\dfrac{3}{8}k = \dfrac{15}{16}$

8. $\dfrac{8}{3}k = \dfrac{15}{16}$

9. $-\dfrac{1}{2} = -\dfrac{3}{8}h$

9. $-\dfrac{1}{2} = -\dfrac{8}{3}h$

Name:
Instructor:

Date:
Section:

10. $-\dfrac{9}{11}=\dfrac{1}{3}y$

10. $\dfrac{-9}{11}=\dfrac{3}{1}y$

11. $\dfrac{2}{9}n=18$

11. $\dfrac{9}{2}n=18$

12. $-\dfrac{7}{5}r=7$

12. $-\dfrac{5}{7}r=7$

Objective 2	Use both the addition and multiplication properties of equality to solve equations containing fractions.

13. $\dfrac{1}{5}n+6=8$
$\dfrac{-\ 8}{2}$

13. $\dfrac{5}{1}n=2$

14. $5+\dfrac{1}{2}p=9$

14. $\dfrac{2}{1}p=4$

15. $-8=\dfrac{5}{2}r+2$

15. $-10=\dfrac{2}{5}r$

16. $0=8+\dfrac{2}{3}t$

16. $-8=\dfrac{3}{2}t$

17. $\dfrac{5}{8}x-10=0$

17. $\dfrac{8}{5}x=10$

18. $\dfrac{1}{5}s - 15 = -10$

18. $\dfrac{5}{1}s = 5$

19. $9 - 5 = \dfrac{1}{2}y - 4$
4

19. $\dfrac{1}{2}y$

20. $0 - 12 = \dfrac{2}{3}k - 6$

20. $K = -9$

21. $3 + \dfrac{1}{3}n = {}^-11 + 5$

21. $n = -27$

22. $\dfrac{3}{5} = \dfrac{7}{3}x - \dfrac{1}{3}$

22. $X = 2/5$

23. $4c - \dfrac{2}{3} = \dfrac{1}{4}$

23. $C = 11/48$

24. $-\dfrac{3}{7} = 18t + \dfrac{3}{14}$

24. _____

Name: 　　　　　　　　　　　　Date:
Instructor: 　　　　　　　　　　Section:

Objective 3　　Solve application problems using equations containing fractions.

In Exercises 25–28, find each person's age using the six problem-solving steps and this expression for approximate systolic blood pressure: $100 + \frac{age}{2}$. *Assume that all the people have normal blood pressure.*

25. A man has a systolic blood pressure of 110. How old is he?

25. _____ 20 _____

26. A woman has a systolic blood pressure of 126. How old is she?

26. $100 + \dfrac{age}{2} = 126$

　　$\underline{-100} \qquad \underline{-100}$

　　$2 \cdot \dfrac{a}{2} = 26 \cdot 2$

　　$\boxed{a = 52}$

27. A man has a systolic blood pressure of 116. How old is he?

27. _____

28. A woman has a systolic blood pressure of 115. How old is she?

28. _____ 30 _____

$100 + \dfrac{x}{2} = 115$

An expression for the recommended weight of an adult is $\frac{11}{2}$ *(height in inches) – 220. In Exercises 29 and 30, find each person's height using this expression and the six problem-solving steps. Assume that all the people are at their recommended weight.*

29. A man weighs 187 pounds. What is his height in inches?

29. _____ 74 in _____

30. A woman weighs 154 pounds. What is her height in inches

30. _____ 60 in _____

135

Name: Date:
Instructor: Section:

4.8   Geometry Applications: Area and Volume

Section 4.8 Objectives
1. Find the area of a triangle.
2. Find the volume of a rectangular solid.
3. Find the volume of a pyramid.

Key Terms: *Answer the following questions about the key terms for Section 4.7*

Volume

1. **Volume** is measure in cubic units,   1._____
   which means
   a. The answer's units will have
      an exponent of 2
   b. The answer's units will have
      an exponent of 3
   c. The answer will be divisible
      by three
   d. The answer will be in square
      inches

Objective 1   **Find the area of a triangle.**

Find the perimeter and area of each triangle.

1.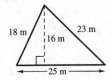

2.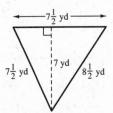

1.

P=_____

A=_____

2.

P=_____

A=_____

Name: Date:
Instructor: Section:

3.

8 in.

8 in.

$11\frac{3}{10}$ in.

3.

P=_____

A=_____

4.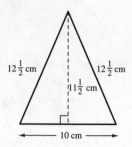

$12\frac{1}{2}$ cm $12\frac{1}{2}$ cm

$11\frac{1}{2}$ cm

10 cm

4.

P=_____

A=_____

5.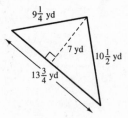

$9\frac{1}{4}$ yd

7 yd $10\frac{1}{2}$ yd

$13\frac{3}{4}$ yd

5.

P=_____

A=_____

6.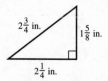

$2\frac{3}{4}$ in. $1\frac{5}{8}$ in.

$2\frac{1}{4}$ in.

6.

P=_____

A=_____

7.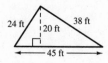

24 ft   20 ft   38 ft

45 ft

7.

P=_____

A=_____

8.

$5\frac{1}{2}$ in. $3\frac{1}{4}$ in.

$4\frac{1}{2}$ in.

8.

P=_____

A=_____

9.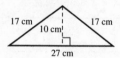

17 cm   17 cm

10 cm

27 cm

9.

P=_____

A=_____

10.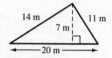

14 m   11 m

7 m

20 m

10.

P=_____

A=_____

Objective 2   **Find the volume of a rectangular solid.**

Find the volume of each solid.

11.

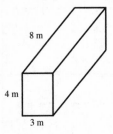

8 m

4 m

3 m

11. *V*=_____

12.

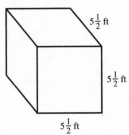

$5\frac{1}{2}$ ft

$5\frac{1}{2}$ ft

$5\frac{1}{2}$ ft

12. *V*=_____

13.

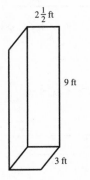

$2\frac{1}{2}$ ft

9 ft

3 ft

13. *V*=_____

14.

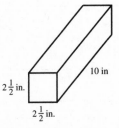

10 in

$2\frac{1}{2}$ in.

$2\frac{1}{2}$ in.

14. *V*=_____

15.

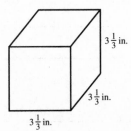

$3\frac{1}{3}$ in.

$3\frac{1}{3}$ in.

$3\frac{1}{3}$ in.

15. *V*=_____

16.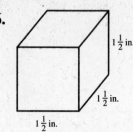

$1\frac{1}{2}$ in.

$1\frac{1}{2}$ in.

$1\frac{1}{2}$ in.

16. *V*=_____

17.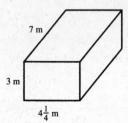

7 m

3 m

$4\frac{1}{4}$ m

17. *V*=_____

Objective 3   **Find the volume of a pyramid.**

18.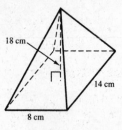

18 cm

14 cm

8 cm

18. *V*=_____

19.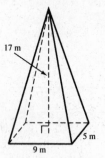

17 m

5 m

9 m

19. *V*=_____

20.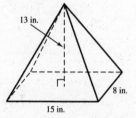

13 in.

8 in.

15 in.

20. *V*=_____

21.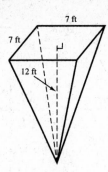

22.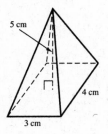

21. _V=_____

22. _V=_____

Chapter 5   RATIONAL NUMBERS: POSITIVE AND NEGATIVE DECIMALS

5.1   Reading and Writing Decimal Numbers

Section 5.1 Objectives
1. Write parts of a whole using decimals.
2. Identify the place value of a digit.
3. Read decimal numbers.
4. Write decimals as fractions or mixed numbers.

Key Terms: *Answer the following questions about the key terms for Section 5.1.*

decimals   decimal point   place value

1. Both of the following are used to   1._____
   represent parts of a whole
   a. **Decimals, ratios**
   b. **Whole numbers, integers**
   c. **Fractions, decimals**
   d. **Integers, fractions**

2. One digit to the right of a **decimal**   2._____ B, _____
   point is called the
   a. Ones place
   b. Tens place
   c. Hundredths place
   d. Tenths place

Objective 1   Write parts of a whole using decimals.

Write the portion of each square that is shaded as a fraction, as a decimal, and in words.

1.(a)

1.(a) $\frac{1}{10}$ one tenth

1.(b)

1.(b) $\frac{63}{100}$ sixty three hundred

Name: _____ Date: _____

Instructor: _____ Section: _____

2.(a)

2.(a) $\dfrac{7}{10}$

2 (b)

2.(b) $\dfrac{29}{100}$

Objective 2 Identify the place value of a digit.

Identify the digit that has the given place value.

3. 30.618
   tens_____
   ones_____
   tenths_____

6. 37.485
   tens_____ ~~7~~ 3
   tenths_____ 4
   hundredths_____ ~~5~~ 8

4. 172.934
   ones_____ ~~4~~ 2
   tenths_____ 9
   tens_____ 7

7. 3792.8154
   thousands_____
   hundreds_____
   thousandths_____

5. 0.32489
   ten-thousandths_____
   tenths_____
   hundredths_____

8. 0.82196
   hundred-thousandths_____ 9
   thousandths_____ 1
   tenths_____ 8

Write the decimal number that has the specified place values.

9. 4 tenths, 7 hundreds, 2 hundredths, 8 tens, 9 ones

9._____

10. 1 hundredth, 3 ten-thousandths, 6 tenths, 2 thousandths, 5 ones

10. 5.6123 ✓

11. 9 thousandths, 3 tenths, 2 ten-thousandths, 4 hundredths

11._____

Name: Date:

Instructor: Section:

12. 0 tens, 3 hundredths, 0 tenths, 9 ones, 7 thousands, 2 hundreds, 1 thousandth

12. *709.031* _____

Objective 3   **Read decimal numbers.**

Tell how to read each decimal in words.

13. 0.7   13._____

14. 0.3   14._*three tenths*_____

15. 0.102   15._____

16. 0.304   16._*three tenths and 4 thousands*_

17. 17.01   17._____

18. 39.05   18._*thirty nine and 5 hundreds*_

Write each decimal in numbers.

19. three hundred eleven thousandths   19._____

20. two hundred forty and three thousandths   20._*.0243*_____

21. three hundred and thirty-nine thousandths   21._____

22. five hundred two ten-
    thousandths   22._____.0502_____

23. nine hundred and five
    hundredths   23._____

24. seventy and twenty
    hundredths   24._____.720_____

Objective 4   **Write decimals as fractions or mixed numbers.**

Write each decimal as a fraction or mixed number in lowest terms

25. 15.4   25._____

26. 0.45   26._____$\frac{45}{100}$_____

27. 0.88   27._____

28. 0.605   28._____$\frac{605}{1000}$_____

29. 7.002   29._____

30. 8.008   30._____$8\frac{8}{100}$_____

5.2   Rounding Decimal Numbers

Section 5.2 Objectives
1. Learn the rules for rounding decimals.
2. Round decimals to any given place.
3. Round money amounts to the nearest cent or nearest dollar.

Key Terms

Answer the following questions about the key terms for Section 5.2

    round   **decimal places**

1. When **rounding** 12,470.635 to the nearest hundredth, what is true?
    a. Since the first digit to be cut is 5, it would round to 12,470.64
    b. Since the first digit to be cut is 5, it would round to 12,470.63
    c. Since the thousands place is 2, it would round to 11,000
    d. Since the hundreds place is 4, it would round to 12,500

1. _____ D. _____

2. To **round** 495.182 to the nearest tenth, what is true?
    a. $495.182 \approx 495.18$
    b. $495.182 \approx 495.19$
    c. $495.182 \approx 495.2$
    d. $495.182 \approx 496$

2. _____ A _____

Objective 1	Learn the rules for rounding decimals.
Objective 2	Round decimals to any given place.

Round each number to the place indicated.

1. 24.7834 to the nearest tenth

1. _____ 24.80 _____

2. 174.745 to the nearest hundredth

2. 174.45

3. 0.85738 to the nearest thousandth

3. 0.8570

4. 74.22573 to the nearest ten-thousandth

4. 74.2257

5. 0.488 to the nearest hundredth

5. 0.490

6. 0.651 to the nearest tenth

6. 0.7

7. 0.953 to the nearest tenth

7. 0.1.00

8. 2.77073 to the nearest thousandth

8. 2.771

9. 672.976 to the nearest tenth

9. 673.

10. 0.08603 to the nearest ten-thousandth

10. 0.0860

11. 36.522 to the nearest one

11. 36.5

12. 8.0708 to the nearest hundredth

12. 8.07

13. 74.000253 to the nearest ten-thousandth

13. _____74.0003_____

14. 0.300581 to the nearest ten-thousandth

14. _0.3006_____

15. 79.8175 to the nearest hundredth

15. _____79.818_____

16. 0.653 to the nearest tenth

16. _0.7_____

17. 27.389 to the nearest one

17. ____27.4_____

18. 0.800545 to the nearest ten-thousandth

18. __0.8005_____ ✓

Objective 3   **Round money amounts to the nearest cent or nearest dollar.**

Round each money amount as indicated.

19. $399.89 to the nearest dollar

19. ____$ 400_____

20. $408.50 to the nearest dollar

20. $409.00_____

21. $0.995 to the nearest cent

21. ____$ 1.00_____

22. $0.08839 to the nearest cent

22. ___$0.09_____

23. $1.0395 to the nearest cent

23. ___$1.04_____

24. $1999.75 to the nearest dollar

24. $ 2000, 00

25. $9999.99 to the nearest dollar

25. $ 1000

26. $9999.095 to the nearest cent

26. $9999.10

27. $275.995 to the nearest cent

27. $ 276.

28. $8899.55 to the nearest dollar

28. $ 8900.00

29. $0.967 to the nearest cent

29. $ 1.00

30. $17.063 to the nearest cent

30. $ 17.06

5.3   Adding and Subtracting Signed Decimal Numbers

Section 5.3 Objectives
1. Add and subtract positive decimals.
2. Add and subtract negative decimals.
3. Estimate the answer when adding or subtracting decimals.

Objective 1   Add and subtract positive decimals.

Find each sum or difference

1. $8 + 0.48 + 3.9$ 1. _____ 12.38 _____

2. $0.813 + 4.6 + 44$ 2. _____ 49.413 _____

3. $80.6 - 0.7$ 3. _____ 799.9 _____

4. $405.63 - 0.39$ 4. _____ 405.24 _____

5. $0.5 - 0.348$ 5. _____ 0.152 _____

6. $0.55 - 0.072$ 6. _____ 6.478 _____

7. $18 - 7.436$ 7. _____ 10.564 _____

8. $80 - 0.08$ 8. _____ 79.92 _____

9. $176.8 + 0.3 + 47.69$ 9. _____ 224.79 _____

10. $0.39 + 0.46 + 0.093 + 0.1$ 10. _____ 1.043 _____

Objective 2   Add and subtract negative decimals.

Find each sum or difference.

11. $-5.03 + 80.6$ **11.** _____ 75.57 _____

12. $0.78 - 4.08$ **12.** _____ -3.3 _____

13. $0.7 - 8.64$ **13.** _____ -7.94 _____

14. $-3 - 7.88$ **14.** _____ -10.88 _____

15. $4.053 - (-62.9)$ **15.** _____ 66.953 _____

16. $-4.005 + 0.38$ **16.** _____ -3.625 _____

17. $-0.43 - 75$ **17.** _____ -75.43 _____

18. $-9.7025 - (6 - 7.8)$ **18.** _____ -7.9025 _____

19. $0.8 - (6 - 8.1)$ **19.** _____ 2.9 _____

20. $39.024 - (-24.1)$ **20.** _____ 63.124 _____

21. $70 + (-0.3951 + 2.6)$ **21.** _____ 72.2049 _____

22. $-350 - (-0.6309 + 0.3)$ **22.** _____ -349.6691 _____

Objective 3   Estimate the answer when adding or subtracting decimals.

Use front end rounding to estimate each answer. Then find the exact answer.

23. 3.92 + 4.008 + 85.3

23.

Estimate: _____ 9.8 _____

Exact: _____ 7.928 _____

24. 8.9 + 5.67 + 92.75

24.

Estimate: _____ 105 _____

Exact: _____ 107.32 _____

25. 12.265 – 27

25.

Estimate: _____ ‒20 _____

Exact: _____ ‒14.735 _____

26. 74.53 – 98

26.

Estimate: _____ ‒30 _____

Exact: _____ ‒23.47 _____

27. –312.8 + 428.66

27.

Estimate: _____ 100 _____

Exact: _____ 155.86 _____

28. –251.3 + 207.81

28.

Estimate: _____ 52208 _____

Exact: _____ ‒43.49 _____

29. Lisa has agreed to work 37.5 hours a week as a sales clerk. So far this week she has worked 18.35 hours. How many more hours must she work?

29.

Estimate: _____

Exact: _____

30. Byron Lee's paycheck stub showed wages of $364.29 at the regular rate of pay and $129.15 at the overtime rate. What were his total wages?

30.

*Estimate:*_____

*Exact:*_____

Name: Date:

Instructor: Section:

5.4   Multiplying Signed Decimal Numbers

Section 5.4 Objectives
1. Multiply positive and negative decimals.
2. Estimate the answer when multiplying decimals.

Objective 1   **Multiply positive and negative decimals.**

Multiply.

1. –3.5(0.7)

1. $\underline{\hspace{1cm} -2.45 \hspace{2cm}}$

2. (65.2)(0.35)

2. $\underline{\hspace{1cm} 22.82 \hspace{2cm}}$

3. (–32.4)(–0.777)

3. $\underline{\hspace{1cm} 25.1748 \hspace{2cm}}$

4. 0.879(–0.638)

4. $\underline{\hspace{1cm} -0.560802 \hspace{2cm}}$

5. –93.1(–2.4)

5. $\underline{\hspace{1cm} 223.44 \hspace{2cm}}$

6. ⁻0.007(0.007)

6. $\underline{\hspace{1cm} -0.000049 \hspace{2cm}}$

7. (0.13)(0.0006)

7. $\underline{\hspace{1cm} 0.000078 \hspace{2cm}}$

8. (–2.987)(–44)

8. $\underline{\hspace{1cm} 131.428 \hspace{2cm}}$

9. $\$742.61$
     $\times \quad 823$

9. $\underline{\hspace{1cm} \$611168.03 \hspace{2cm}}$

10. $42.71   10._____ $13069.26 _____
     ×   306

11. (−3.724)(−61)   11._____ 227.164 _____

12. $689.42   12._____ 164827.12 _____
     ×   236

13. 0.793(−0.627)   13._____ $ 164827.12 _____

14.   82.1   14._____ 76.353 _____
     × 0.93

Objective 2   **Estimate the answer when multiplying decimals.**

First, use front end rounding to estimate each answer. Then find the exact answer.

15. (10.39)(5.03)   **15.**

                         *Estimate:*_____

                         *Exact:*_____

16. (−3.975)(−34)   **16.**

                         *Estimate:*_____

                         *Exact:*_____

17. 9.32(3.2)   **17.**

                         *Estimate:*_____

                         *Exact:*_____

18.   7.89   **18.**
     × 5.2
                         *Estimate:*_____

                         *Exact:*_____

19.   39.53
       × 21.32

19.

Estimate:_____

Exact:_____

20. 59.6(−18.3)

20.

Estimate:_____

Exact:_____

21. (15.67)(4.52)

21.

Estimate:_____

Exact:_____

22. (−16.525)(−28)

22.

Estimate:_____

Exact:_____

23.   7.08
       ×   61

23.

Estimate:_____

Exact:_____

24.   8.209
       × 6.59

24.

Estimate:_____

Exact:_____

Solve each application problem. Round money answers to the nearest cent when necessary.

25. Jeanie's time card shows 47.6 hours at $12.45 per hour. What are her gross earnings?

25._____

26. How much will Ms. Quinn pay for 3.75 pounds of cheese that costs $1.29 per pound?

26._____

27. Brent filled the tank of his car with regular unleaded gas that costs $2.949 per gallon. If he bought 17.503 gallons, how much did he pay?

27._____

28. Paper for the copy machine at a print shop costs $0.012 per sheet. How much will the shop pay for 6750 sheets?

28._____

29. Paul Nauman is a real estate broker who helps people sell their homes. His fee is 0.08 times the price of the home. What was his fee for selling a $234,500 home?

29._____

30. Charles worked 53.7 hours over the last two weeks. He earned $13.72 per hour. How much did he make?

30._____

5.5   Dividing Signed Decimal Numbers

Section 5.5 Objectives
1. Divide a decimal by an integer.
2. Divide a number by a decimal.
3. Estimate the answer when dividing decimals.
4. Use the order of operations with decimals

Key Terms: *Answer the following questions about the key terms for Section 5.5.*

repeating decimal

1. The number $1.\overline{3}$ means   1._____
   a. The number is undefined
   b. The 1 is a fraction
   c. The 3 repeats forever
   d. The 3 is an estimate

Objective 1   Divide a decimal by an integer.

Divide.

1. $89.2 \div (-5)$ 1._____

2. $-30.6 \div 6$ 2._____

3. $4\overline{)86.5}$ 3._____

4. $5\overline{)3.092}$ 4._____

5. $-125.85 \div (-15)$ 5._____

6. $\dfrac{9.76}{8}$ 6._____

7. $-0.5324 \div (-4)$ 7._____

8. $-375.84 \div 12$ 8._____

Objective 2   **Divide a number by a decimal.**

Divide. Round quotients to the nearest hundredth when necessary.

9. $0.04\overline{)6.8035}$ 9._____

10. $0.002\overline{)37}$ 10._____

11. $-4.6 \div 0.075$ 11._____

12. $\dfrac{8.6}{1.1}$ 12._____

13. $2.5\overline{)77}$ 13._____

14. $\dfrac{-2.79}{0.21}$ 14._____

15. $0.035\overline{)643.92}$ 15._____

16. $\dfrac{-3.2}{-0.007}$ 16._____

Solve each application problem. Round money answers to the nearest cent when necessary.

17. The Riverton School purchased   17._____
    13 ink cartridges for the
    computer lab printers for
    $228.67. How much did they
    pay for each cartridge?

18. The grocery store has a special   18._____
    price on chips: 3 bags for
    $5.67. How much is one bag of
    chips?

19. Cindy Hein bought 1.7 meters   19._____
    of curtain fabric for $13.52.
    How much did he pay per
    meter?

20. Tissues are on sale at four   20. _____
    boxes for $5.28, or you can
    purchase individual boxes for
    $1.55. How much will you
    save per box if you buy four
    boxes?

Objective 3   **Estimate the answer when dividing decimals.**

Decide whether each answer is reasonable or unreasonable by using front end rounding to estimate the answer.

If the exact answer is not reasonable, find and correct the error.

21. $89.17 \div 1.85 = 4.82$ 21.

                                       *Estimate:*_____

22. $289.92 \div 6 = 48.32$

22.

Estimate:_____

23. $3.182 \div 0.86 = 3.7$

23.

Estimate:_____

24. $96 \div 17.472 = 182$

24.

Estimate:_____

Objective 4   **Use the order of operations with decimals.**

Simplify by using the order of operations

25. $8.3 - 2.4 + 6.7^2$

25._____

26. $7.25 - 2.07(0.96 - 8.95)$

26._____

27. $-3.45 - 2.3(8.6) \div 0.5$

27._____

28. $24.1 + 12.6 \div 6.3(-2.75)$

28._____

29. $0.5 + (-1.87 + 0.22) \div 0.002(0.5)$

29._____

30. $-7.84 - 5.2(8.9 \div 2) + 6.93$

30._____

5.6   Fractions and Decimals

Section 5.6 Objectives
1. Write fractions as equivalent decimals.
2. Compare the size of fractions and decimals.

Objective 1   Write fractions as equivalent decimals.

Write each fraction or mixed number as a decimal. Round to the nearest thousandth when necessary.

1. $1\dfrac{3}{8}$ 1._____

2. $\dfrac{7}{20}$ 2._____

3. $1\dfrac{3}{5}$ 3._____

4. $\dfrac{7}{9}$ 4._____

5. $\dfrac{1}{20}$ 5._____

6. $3\dfrac{7}{10}$ 6._____

7. $\dfrac{5}{8}$ 7._____

8. $3\dfrac{8}{9}$ **8.**_____

9. $10\dfrac{4}{7}$ **9.**_____

10. $15\dfrac{1}{2}$ **10.**_____

11. $24\dfrac{3}{4}$ **11.**_____

12. $5\dfrac{2}{3}$ **12.**_____

13. $\dfrac{9}{11}$ **13.**_____

14. $2\dfrac{1}{6}$ **14.**_____

15. $\dfrac{3}{20}$ **15.**_____

16. $17\dfrac{1}{4}$ **16.**_____

17. $4\dfrac{2}{9}$ **17.**_____

18. $5\dfrac{1}{3}$ **18.**_____

| Objective 2 | **Compare the size of fractions and decimals.** |

Arrange each group of numbers in order from smallest to largest.

19. 0.8, 0.804, 0.8039   **19.**_____

20. 5.49, 5.049, 5.5, 5.501   **20.**_____

21. 0.8005, 0.85, 0.8   **21.**_____

22. 13.99, 13.87, 13.5, 14.001   **22.**_____

23.   6.704, 6.0069, 6.69, 6.7   **23.**_____

24. 1.0875, $1\dfrac{7}{8}$, $1\dfrac{3}{4}$, 0.9   **24.**_____

25. $\dfrac{1}{3}$, $\dfrac{4}{7}$, 0.5, 0.35   **25.**_____

26. $\dfrac{1}{9}$, $\dfrac{10}{11}$, 0.09, 0.109   **26.**_____

Solve each application problem.

27. A patient in the hospital is supposed to get 7.5 milligrams of medicine. She was actually given 7.055 milligrams. Did she get too much or too little medicine? What was the difference?

27._____

28. The average automatic teller transaction takes 2.7 minutes. Elizabeth's transaction took $2\frac{7}{8}$ minutes. Was Elizabeth's time longer or shorter than the average? By how much?

28._____

29. A ski resort owner hoped a snowstorm would drop $5\frac{1}{2}$ inches of snow. The newspaper said the area received 5.2 inches. Was that more or less than the owner hoped for? By how much?

29._____

30. The label on the bottle of medicine says that each capsule contains 0.003 gram of sodium. When checked, each capsule had 0.030 gram of sodium. Was there too much or too little sodium? What was the difference?

30._____

5.7 Problem Solving with Statistics: Mean, Median, Mode, and Variability

Section 5.7 Objectives
1. Find the mean of a list of numbers.
2. Find a weighted mean.
3. Find the median.
4. Find the mode.
5. Evaluate the variability of a set of data by finding the range of values.

Key Terms: *Answer the following questions about the key terms for Section 5.7*

> **mean**   **weighted mean**   **median**
>
> **mode**   **variability**

1. Another word for **mean** is
    a. Average
    b. middle
    c. most common
    d. range of values

1._____

2. To find a **weighted mean** in the following table,

Scores on a Quiz	Frequencies
10	1
9	2
7	6
5	1

    a. multiply each score by 2, add them, and divide by 10
    b. divide the total by 20
    c. multiply each score by its frequency, add the products, divide by 10
    d. see what the most frequent score was

2._____

3. In the same table of Quiz Scores above, the **median** score would be
    a. 9
    b. 5
    c. 7
    d. 8

3._____

4. In the same table of Quiz Scores above, the **mode** is 7 because

    a. It is the average score

    b. It is the middle score

    c. It is the most frequent score

    d. It is a passing score

4. _____

5. In the table below are quiz scores for Quiz 1 and Quiz 2. Which quiz had greater variability?

5. _____

Scores on Quiz 1	Frequencies	Scores on Quiz 2	Frequencies
10	1	9	3
9	2	7	5
7	6	6	1
5	1	5	1

a. Quiz 1

b. Quiz 2

c. Quiz 1 and Quiz 2 had the same variability

Objective 1   **Find the mean of a list of numbers.**

**Find the mean for each list of numbers. Round answers to the nearest tenth when necessary.**

1. Monthly long distance phone bills: $32.50, $47.22, $13.25, $56.40, $82.66, $19.71, $12.88, $51.25, $18.52

1. _____

2. Quiz scores: 21, 25, 18, 20, 13, 19, 22, 24, 23, 25

2. _____

3. Annual salaries: $42,900, $37,600, $28,950, $48,500, $50,000, $53,275, $41,000

3. _____

4. Ages of patients at the hospital (in years): 38, 27, 68, 5, 88, 92, 39, 27, 62

4. _____

5. Numbers of people attending high school football games: 760, 800, 1225, 930, 839, 1050, 346

5. _____

6. Final exam scores: 93, 52, 58, 79, 83, 32, 69, 39, 68, 77

6. _____

Solve each application problem.

7. The Fancy Framing Store sold prints at the following prices: $392, $260, $899, $123, $275.
   Find the mean price for a print.

7. _____

8. In one evening, a waitress collected the following amounts in tips: $7.75, $10.50, $12.00, $6.50, $13.00, $24.00, $11.75, $4.50.
   Find the mean tip amount.

8. _____

Objective 2   **Find a weighted mean.**

Find the weighted mean. Round answers to the nearest tenth when necessary.

9.
Hours worked	Frequency
20	4
22	3
25	1
27	6
32	1

9. _____

10.

Students per class	Frequency
18	2
19	3
20	4
22	2
24	3
25	2

10._____

11.

Quiz Scores	Frequency
4	5
5	6
6	2
7	8
9	3

11._____

12.

Credits per Student	Frequency
8	3
13	5
16	6
17	2
19	1

12._____

Objective 3   **Find the median.**

Find the median for each list of numbers.

13. Number of liters of pop sold each day:
27, 30, 16, 25, 28, 10, 29, 37

13._____

14. Number of pieces of mail delivered each day: 7, 12, 3, 0, 2, 5, 3, 7, 8

14._____

Objective 4   **Find the mode.**

Find the mode or modes for each list of numbers.

15. Ages of starting teachers (in years): 23, 25, 28, 24, 25, 36, 22

15._____

16. Monthly electric bills: $32, $27, $26, $17, $27, $26, $33, $32, $19, $27, $43

16._____

17. Number of service calls received each hour: 8, 9, 3, 2, 7, 6, 9, 5, 3

17._____

18. Number of hawks spotted each hour: 47, 32, 5, 16, 35, 24, 18, 40

18._____

19. Number of defective computer disks: 10, 13, 9, 8, 15, 13, 12, 7, 11

19._____

20. Hours worked per week: 41, 37, 42, 50, 40, 38, 42, 36, 40, 35

20._____

21. Weekly salaries: $800, $750, $395, $700, $460, $685

21._____

22. Miles per gallon for cars: 24, 38, 18, 17, 32, 24, 23, 25, 18

22._____

Find the mean, median, and mode(s) for each list of numbers. Round answers to the nearest tenth when necessary.

23 Students per class: 30, 26, 32, 20, 33,   23._____
      27, 30, 32, 27, 26, 30, 24

24.   Heights of basketball players (in   24._____
      inches): 72, 74, 73, 70, 78, 79, 77,
      76, 73, 75, 73, 72, 77

Find the GPA (grade point average) for students earning the following grades. Assume A = 4, B = 3, C = 2, D = 1, and F = 0. Round answers to the nearest hundredth

25.	Course	Credits	Grade		25.	GPA
	Health	3	B			
	Mathematics	4	A			
	History	2	C			
	English	4	B			

26.	Course	Credits	Grade		26.	GPA
	Chemistry	3	D			
	Biology	3	C			
	Psychology	2	A			
	Mathematics	4	B			
	Theater	2	A			

Objective 5   **Evaluate the variability of a set of data by finding the range of values.**

Find the range for each set of data. Which set has greater variability?

27. Student's test scores: 27._____
      Student A: 90, 83, 52, 73, 69
      Student B: 80, 95, 42, 77, 68

28. Monthly salaries:
Company X: $3200, $3500, $3800,
$4100, $2800
Company Y: $4500, $4200, $3900,
$5200, $3700
Company Z: $5700, $4750, $4800,
$6200, $5000

28._____

29. Daily high temperatures (in °F)
Week J: 48, 52, 61, 50, 55, 57, 66
Week K: 73, 79, 83, 80, 75, 82, 69

29._____

30. Number of samples taken each day:
Lab Q: 8, 6, 4, 3, 10, 5
Lab R: 2, 1, 12, 7, 6, 5

30._____

Name: _____ Date: _____

Instructor: _____ Section: _____

5.8 Geometry Applications: Pythagorean Theorem and Square Roots

Section 5.8 Objectives
1. Find square roots using the square root key on a calculator.
2. Find the unknown length in a right triangle.
3. Solve application problems involving right triangles.

Key Terms: *Answer the following questions about the key terms for Section 5.8.*

square root   **hypotenuse**

1. $3^2 = \sqrt{9}$

      a. True

      b. False

1. _____

2. The difference between a legs in a right triangle and the **hypotenuse** is

      a. the legs are always longer

      b. the hypotenuse is on the bottom

      c. the hypotenuse is the longest side in a right triangle

      d. the angles formed by the intersection of the hypotenuse and the legs are 90°

2. _____

Objective 1   **Find square roots using the square root key on a calculator.**

Find each square root. Starting with Exercise 5, find the square root using a calculator. Round answers to the nearest thousandth when necessary.

1. $\sqrt{10,000}$

1. _____

2. $\sqrt{121}$

2. _____

3. $\sqrt{49}$

3. _____

4. $\sqrt{36}\,5$

4. _____

5. $\sqrt{13}$

5. _____

6. $\sqrt{24}$

6. _____

7. $\sqrt{90}$

7. _____

8. $\sqrt{201}$

8. _____

9. $\sqrt{69}$

9. _____

10. $\sqrt{324}$

10. _____

11. $\sqrt{150}$

11. _____

12. $\sqrt{78}$

12. _____

13. $\sqrt{3000}$

13. _____

14. $\sqrt{676}$

14. _____

Objective 2   Find the unknown length in a right triangle.

Find the unknown length in each right triangle. Use a calculator to find square roots.
Round your answers to the nearest tenth when necessary

15.

15._____

16.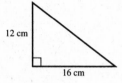

16._____

17.

17._____

18.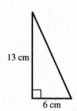

18._____

19.

19._____

20.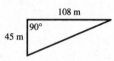

20._____

21.
1.5 in.
3.5 in.

21._____

22.
26.7 ft
38.4 ft

22._____

23.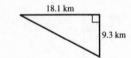
12 cm
5 cm
?

23._____

24.
9 in.
15 in.

24._____

25.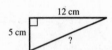
18.1 km
9.3 km

25._____

26.
19 m
23 m
90°

26._____

Objective 3   **Solve application problems involving right triangles.**

Solve each application problem. Round your answers to the nearest tenth when necessary

27. Find the length of the ladder.

27._____

28. Find the height of the flag pole.

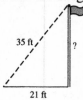

28._____

29. How high is the airplane off the ground?

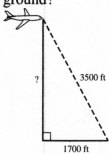

29._____

30. Find the length of this loading ramp.

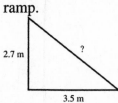

30._____

5.9   Problem Solving: Equations Containing Decimals

Section 5.9 Objectives
1. Solve equations containing decimals using the addition property of equality.
2. Solve equations containing decimals using the division property of equality.
3. Solve equations containing decimals using both properties of equality.
4. Solve application problems involving equations with decimals.

Objective 1   Solve equations containing decimals using the addition property of equality.

Solve each equation.

1. $13.4 + h = 7$

$13.4 + h = 7$
$-7 = h$
1. $\underline{\;\;(6.4 = h)\;\;}$

2. $^-0.55 + y = 0$

$-0.55 + y = 0$
2. $\underline{\;\;(0.55 = y)\;\;}$

3. $c - 3.8 = {}^-15.6$

$c - 3.8 = {}^-15.6$
$+3.8 \quad +3.8$
3. $\underline{\qquad\qquad}$
$(c = {}^-11.8)$

4. $^-0.3 = k - 0.9$

$-0.3 = k - .09$
$+.3$
4. $\underline{\qquad\qquad}$
$(k = 1.2)$

5. $^-30.4 + n = {}^-35$

$-30.4 + n = -35$
$+30.4$
5. $\underline{\qquad\qquad}$
$(n = -4.6)$

6. $g - 7 = 8.04$

$g - 7 = 8.04$
7
6. $\underline{\qquad\qquad}$
$(g = 15.04)$

7. $0 = b - 0.003$

$0 = b - 0.003$
7. $\underline{\;(0.003 = b)\;}$

8. $^-7.1 = 0.32 + m$

$-7.1 = 0.32 + m$
$-3.2 + .32$
8. $\underline{\qquad\qquad}$
$(-10.3 = m)$

Objective 2 Solve equations containing decimals using the division property of equality.

Solve each equation.

9. $^-2y = \dfrac{-0.76}{2}$

9. ___$y = -0.38$___

10. $^-8.9c = \dfrac{0}{-8.6}$

10. ___$C = -0$___

11. $7.3w = \dfrac{^-58.4}{7.3}$

11. ___$W = -8$___

12. $^-6.6t = \dfrac{^-5.28}{-6.6}$

12. ___$t = 0.8$___

13. $0 = 11.7m$
$\overline{11.7}$

13. ___$m = 0$___

14. $^-8.7 = \dfrac{1.5t}{1.5}$

14. ___$t = -5.8$___

Objective 3 Solve equations containing decimals using both properties of equality.

Solve each equation.

15. $^-4.5 = 1.2m + 0.18$
$\quad -.18 \qquad -.18$
$\overline{\quad -4.68 = 1.2m}$
$\qquad\qquad 1.2$

15. ___$m = -3.9$___

16. $7w - 7.3 = ^-7.3 + 2w$
$\quad -2w \qquad\qquad 2w$
$\overline{\quad 5w - 7.3 = ^-7.3}$
$\qquad 7.3 \qquad +7.3$
$\overline{\qquad\quad 5w =}$

16. _____

17. $3.7h + 7 = 1.3h + 13$ 17._____

18. $8.5x + 0.27 = -5$ 18._____

19. $0.7 = 0.3y + 4.9$ 19._____

20. $8.4h - 1.8 = 2.4$ 20._____

21. $0.6x + 4.98 = x - 6.78$ 21._____

22. $2c + 11 = 5c + 9.35$ 22._____

23. $-2.01 = 9.6y - 0.006$ 23._____

24. $^-11.2 - 0.7p = 0.9p + 5.6$ 24._____

25. $9r + 8.34 = {}^-2.36 + 7r$ **25.**_____

26. ${}^-3.25 = 7.61z - 3.25$ **26.**_____

Objective 4   **Solve application problems involving equations with decimals.**

Solve each application problem using the six problem-solving steps

27. For a 30-pound child, an adult dose   **27.**_____
of medication should be multiplied
by 0.2. If the child's dose of a
cough suppressant is 10 milliliters,
find the adult dose.

28. An air compressor can be rented for   **28.**_____
$38.95 for the first three hours, and
$8.50 for each additional hour.
William's rental charge was
$64.45. How many hours did he
rent the compressor?

29. When using a particular calling   **29.**_____
card, the cost is $1.15 per minute
plus a $0.75 service charge per call.
If Julie was billed $20.30 for one
call, how long did the call last?

30. Most adult medication doses are for   **30.**_____
a person weighing 150 pounds. For
a 45-pound child, the adult dose
should be multiplied by 0.3. If the
child's dose of a decongestant is 9
milligrams, what is the adult dose?

5.10   Geometry Applications: Circles, Cylinders, and Surface Area

Section 5.10 Objectives
1. Find the radius and diameter of a circle.
2. Find the circumference of a circle.
3. Find the area of a circle.
4. Find the volume of a cylinder.
5. Find the surface area of a rectangular solid.
6. Find the surface area of a cylinder.

Key Terms: *Answer the following questions about the key terms for Section 5.10.*

circle   radius   diameter

circumference π (pi)   surface area

1. The most important characteristic   1._____
    of a **circle** is
    a. All points are the same
        distance from the center
    b. It has sides of equal length
    c. All angles measure 90°
    d. It has a base and a height

2. The difference between the **radius**   2._____
    and the **diameter** is
    a. The **radius** is longer than
        the **diameter**
    b. The **radius** is half as long as
        the **diameter**
    c. The **radius** and the
        diameter are the same
        length
    d. The **radius** is equal to π
        (pi), the **diameter** is not

3. **Circumference** in circles is a   3._____
    similar measure to
    a. area in rectangles
    b. hypotenuse in triangles
    c. perimeter in rectangles
    d. side in squares

4. We use **π (pi)** to find the   4._____
    a. **Diameter** of a circle
    b. **Radius** of a circle
    c. **Square root** of a circle
    d. **Circumference** of a circle

5. The **surface area** of a shipping   5._____
carton is measured in
    a. Cubic units (x^3)
    b. Units of volume
    c. Units of weight
    d. Square units (x^2)

Objective 1   **Find the radius and diameter of a circle.**

Find the unknown length of the diameter or radius in each circle.

1.   1._____

60 ft

2.   2._____

13 cm

3.   3._____

33 yd

4.   4._____

7.5 m

Objective 2   **Find the circumference of a circle.**

5.

160 mm

5.

6. *diameter = 0.8 km*

6. .

7.

11 in.

7.

8. *radius = 7.8 m*

8._____

Objective 3 Find the area of a circle.

Find the circumference of each circle. Use 3.14 as the approximate value for π. Round your answers to the nearest tenth

Find the area of each circle. Use 3.14 as the approximate value for π. Round your answers to the nearest tenth.

9.

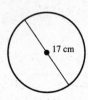

9._____

10. diameter = 9.4 cm

10._____

11.

11._____

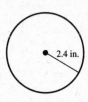

12. radius = 0.73 in.

12._____

13.. $d = 19$ cm

13._____

14. $d = 9.78$ mm

14._____

15. $d = 6\frac{1}{2}$ ft

15._____

16. $d = 51$ in.

16._____

17. $r = 4.5\ m$

17._____

18. $r = 12$ yd

18._____

Objective 4   Find the volume of a cylinder.

Find the volume of each cylinder. Use 3.14 as the approximate value for π. Round your answers to the nearest tenth

19.

19._____

20.

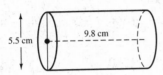

20._____

Objective 5   Find the surface area of a rectangular solid.

Find the surface area of each rectangular solid. Round your answers to the nearest tenth.

21.

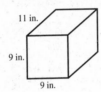

21._____

22.

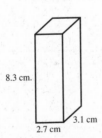

22_____

Objective 6   Find the surface area of a cylinder.

Find the surface area of each cylinder. Use 3.14 as the approximate value for π. Round your answers to the nearest tenth

23.

23._____

24.

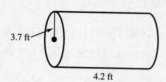

3.7 ft

4.2 ft

24._____

25.

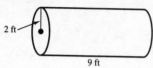

12.6 in.

1.7 in.

1.5 in.

25._____

26.

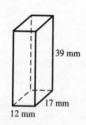

2 ft

9 ft

26._____

27.

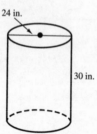

39 mm

17 mm

12 mm

27._____

28.

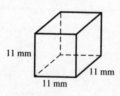

24 in.

30 in.

28._____

29.

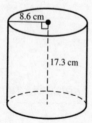

11 mm

11 mm

11 mm

29._____

30.

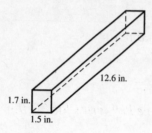

8.6 cm

17.3 cm

30._____

Chapter 6 RATIO, PROPORTION, AND LINE/ANGLE/TRIANGLE RELATIONSHIPS

6.1   Ratios

Section 6.1 Objectives
1. Write ratios as fractions.
2. Solve ratio problems involving decimals or mixed numbers.
3. Solve ratio problems after converting units.

Key Terms *Answer the following questions about the key terms for Section 6.1.*

Ratio

1. To write the following **ratio, 8 m to**   1. _____A._____
    12m as a fraction, the numerator will
    be
    a. 8
    b. 12

Objective 1   **Write ratios as fractions.**

Write each ratio as a fraction in lowest terms.

1. 3 to 11   1. ___$\frac{3}{11}$_____

2. 11 to 17   2. ___$\frac{11}{17}$_____

3. $300 to $100   3. ___$\frac{300}{100}$_____

4. 45¢ to 9¢   4. ___$\frac{45}{9}$_____

5. 20 minutes to 80 minutes   5. ___$\frac{20}{80}$_____

6. 5 pounds to 30 pounds   6. ___$\frac{5}{30}$_____

7. 90 miles to 70 miles

7. _____ $\frac{90}{70}$ _____

8. 150 people to 250 people

8. _____ $\frac{150}{250}$ _____

9. 18 hours to 24 hours

9. _____ $\frac{18}{24}$ _____

10. 55 books to 35 books

10. _____ $\frac{55}{35}$ _____

Objective 2   Solve ratio problems involving decimals or mixed numbers.

Write each ratio as a ratio of whole numbers in lowest terms.

11. $2\frac{1}{2}$ to 5

11. _____ $\frac{5}{5} = 1$ _____

12. $6.50 to $2.50

12. _____ $\frac{6.50}{2.50} = 3$ _____

13. $0.12 to $0.08

13. _____ $\frac{.12}{.08}$ _____

14. 17 to $3\frac{1}{2}$

14. _____ $\frac{17}{7}$ _____

15. $7\frac{1}{2}$ hours to $1\frac{1}{4}$ hours

15. _____ $\frac{15}{5} = 3$ _____

16. $1\frac{1}{2}$ inches to $\frac{3}{8}$ inch

16. _____ $\frac{4}{3/8}$ _____

192

17. $4\dfrac{1}{3}$ cups to $3\dfrac{2}{3}$ cups

17. $\dfrac{13}{11}$

18. $2\dfrac{1}{2}$ feet to 8 feet

18. $\dfrac{5}{8}$

Objective 3 Solve ratio problems after converting units.

Write each ratio as a fraction in lowest terms.

19. 8 inches to 3 feet

19. $\dfrac{8}{3}$

20. 4 feet to 4 yards

20. $\dfrac{4}{4} = 1$

21. 12 minutes to 1 hour

21. $\dfrac{12}{1} = 12$

22. 6 quarts to 8 pints

22. $\dfrac{6}{8} = \dfrac{3}{4}$

23. 20 hours to 2 days

23. $\dfrac{20}{2} = \dfrac{10}{1} = 10$

24. 4 pounds to 10 ounces

24. $\dfrac{4}{10}$

25. 4 gallons to 6 quarts

25. $\dfrac{4}{6} = \dfrac{2}{3}$

26. 5 cups to 6 pints

26. $\dfrac{5}{6}$

27. 5 feet to 6 yards

27. $\dfrac{5}{6}$

28. 5 feet to 20 inches

28. $\dfrac{5}{20} = \dfrac{1}{4}$

29. 7 cups to 2 pints

29. $\dfrac{7}{2}$

30. 4 minutes to 120 seconds

30. $\dfrac{4}{120} = \dfrac{1}{30}$

6.2   Rates

Section 6.2 Objectives
1. Write rates as fractions.
2. Find unit rates.
3. Find the best buy based on cost per unit.

Key Terms *Answer the following questions about the key terms for Section 6.2.*
        rate   **unit rate**   **cost per unit**

1. The **unit rate has**
   a. 1 as the numerator
   b. 1 as the denominator

1. _____ A _____

2. The **rate** you will pay for one gallon of gas is called the
   a. **Unit rate**
   b. **Ratio** in lowest terms
   c. Fractional **rate**
   d. **Cost per unit**

2. _____ D _____

3. One difference between **rate** and **ratio** is
   a. A **rate** compares measurements with the same type of units
   b. A **rate** compares measurements with different types of units
   c. **Rates** are used for money, but **ratios** aren't
   d. **Rates** may be written three ways, but a **ratio** must be written as a fraction.

3. _____

Objective 1   **Write rates as fractions.**

Write each rate as a fraction in lowest terms.

1. 39 servings for 13 people

1. $\dfrac{39}{13} = \dfrac{3}{1}$

2. 200 miles in 16 hours

2. _____

3. $18 for 20 cards

3. $\dfrac{18}{20} = \dfrac{9}{10}$

4. 15 people for 45 coats

4. _____

5. 28 feet in 72 seconds

5. $\dfrac{28}{72} = \dfrac{14}{36} = \dfrac{7}{18}$

6. 14 cups for 6 people

6. _____

7. 63 miles on 3 gallons

7. $\dfrac{63}{3} = \dfrac{21}{1} = 21$

8. $208 for 39 visits

8. _____

9. 30 packages in 18 minutes

9. $\dfrac{30}{18} = \dfrac{5}{3}$

10. 6 teachers for 57 students

10. _____

11. 1380 bushels on 50 acres

11. $\dfrac{1380}{50} = \dfrac{138}{5}$

12. $8 for 20 packages

12. _____

Objective 2   **Find unit rates.**

Find each unit rate.

13. $883.20 for 6 days

13. $\dfrac{883.20}{6} = 13.86$

14. 10.5 pounds for 6 people

14._____

15. $92 in 4 hours

15. $\dfrac{92}{4} = 23$_____

16. 38 cars from 19 families

16._____

17. $108 for 12 visits

17. $\dfrac{108}{12} = 9$_____

18. $101.75 for 11 hours

18._____

19. 550 miles in 10 hours

19. $\dfrac{550}{10} = 55$_____

20. 238 miles on 8.5 gallons

20._____

Solve each application problem.

21. Victoria's stew recipe uses four pounds of beef to feed 8 people. Give the rate in pounds per person.

21. 4^4 to $8 = .5$_____

22. Madeline works 12 hours to earn $164.88. What is her pay rate per hour?

22. $12/164.88 = \$3.74$_____

23. Paula drove 175 miles in 3.5 hours. What was her rate per hour?

23._____

24. Tristan lost 6.8 pounds in five weeks. What was his rate of loss in pounds per week?

24. $6.8/5 = 1.36/wk.$_____

Objective 3   Find the best buy based on cost per unit.

Find the best buy (based on the cost per unit) for each item.

25. Hair conditioner
    8 ounces for $1.29
    12 ounces for $1.99

25. _____ 6.20 = 6.03 _____

26. Canned fruit
    3 cans for $3.45
    4 cans for $4.65
    7 cans for $8.00

26. _____ .86 = .86 . 87 _____

27. Crackers
    12 ounces for $2.59
    16 ounces for $3.39
    20 ounces for $4.39

27. _____

28. Frozen vegetables
    8 ounces for $1.79
    16 ounces for $3.59
    24 ounces for $5.29
    36 ounces for $8.19

28. _____ 4.46 = 4.45 = 4.53 = 4.39 _____

Solve each application problem.

29. Three brands of peanut butter are available. Brand B is priced at $1.99 for 20 ounces. Brand H is $2.89 for 24 ounces and Brand R is $3.59 for 28 ounces. You have a coupon for 45¢ off Brand H and a coupon for 75¢ off Brand R. Which peanut butter is the best buy based on cost per unit?

29. _____ 10.05 = 7.85 = 7.04 _____

Brand B = 20 for 1.99 ≈ 10.05

Brand H = 24 for 2.89 − .45

Brand R = 28 for 3.59 − .75

Brand B

30. Two brands of diapers are available. Brand A is on special at two packages of 22 diapers each for $12. Brand T is priced at $13.95 per package of 45 diapers. You have a coupon for 40¢ off one package of Brand A and a coupon for $1.70 off one package of Brand T. How can you get the best buy on one package of diapers?

30._____

Name: Date:
Instructor: Section:

6.3   Proportions

> **Section 6.3 Objectives**
> 1. Write proportions
> 2. Determine whether proportions are true or false.
> 3. Find the unknown number in a proportion.

Key Terms *Answer the following questions about the key terms for Section 6.3.*

proportion   cross products

1. Determine whether the following   1._____
    proportions are equivalent.
    $10 is to 2 hours as $50 is to 10 hours.
        a. Equivalent
        b. Not equivalent

2. If **cross products** are unequal, the   2._____
    proportion is
        a. True
        b. False

Objective 1   Write proportions.

Write each proportion.

1. (a) $12 is to 18 boxes as $18 is to 27   **1.** $\dfrac{12}{18} = \dfrac{18}{27}$
    boxes.   **(a)**_____

    (b)   15 inches is to 40 inches
    as 3 inches is to 8 inches.   **(b)** $\dfrac{15}{40} = \dfrac{3}{8}$
                                                   _____

2. (a)   400 children is to 700 adults   **2.** $\dfrac{400}{700} = \dfrac{16}{28}$
    as 16 children is to 28 adults.   **(a)**_____

    (b)   48 hours is to 8 hours
    as 66 hours is to 11 hours.   **(b)** $\dfrac{48}{8} = \dfrac{66}{11}$
                                             _____

Objective 2   **Determine whether proportions are true or false.**

Determine whether each proportion is true or false by writing the ratios in lowest terms. Show the simplified ratios and then write true or false.

3. $\dfrac{6}{27} = \dfrac{2}{9}$ 3.   54 = 54

4. $\dfrac{4}{5} = \dfrac{28}{35}$ 4.   140 = 140

5. $\dfrac{25}{40} = \dfrac{5}{6}$ 5.   150 ≠ 200

6. $\dfrac{5}{8} = \dfrac{30}{56}$ 6.   280 ≠ 240

Use cross products to determine whether each proportion is true or false. Show the cross products and then write true or false.

7. $\dfrac{3}{5} = \dfrac{15}{25}$ 7.   75 = 75

8. $\dfrac{20}{42} = \dfrac{4}{7}$ 8.   140 ≠ 168

9. $\dfrac{7}{42} = \dfrac{15}{84}$ 9.

10. $\dfrac{36}{81} = \dfrac{52}{117}$ 10.   612 ≠ 4212

11. $\dfrac{4.8}{7.8} = \dfrac{9.6}{14.3}$

11. _____

12. $\dfrac{24}{11} = \dfrac{3}{1\frac{3}{8}}$

12. $\underline{\quad 264 \neq 33 \quad}$

Objective 3   **Find the unknown number in a proportion.**

Find the unknown number in each proportion. Round your answers to hundredths when necessary. Check your answers by finding cross products.

13. $\dfrac{1}{5} = \dfrac{x}{60}$

13. $5X = 60 \qquad \boxed{X = 12}$

14. $\dfrac{x}{5} = \dfrac{27}{45}$

14. $45X = \dfrac{(27)(5)}{45} \quad X = 3$

15. $\dfrac{40}{35} = \dfrac{8}{x}$

15. _____

16. $\dfrac{2\frac{1}{4}}{3} = \dfrac{x}{4}$

16. $3X = \dfrac{(5)(4)}{3} \quad X = 6.\overline{6}$

17. $\dfrac{x}{11} = \dfrac{45}{5}$

17. _____

18. $\dfrac{24}{x} = \dfrac{6}{21}$

18. $6X = (24)(21) \qquad X = 84$

19. $\dfrac{37}{7} = \dfrac{x}{14}$

19._____

20. $\dfrac{16}{28} = \dfrac{20}{x}$

20._____ $16x = (20)(28)$ $X = 35$

21. $\dfrac{x}{3} = \dfrac{2\frac{2}{3}}{2}$

21._____

22. $\dfrac{0.08}{x} = \dfrac{0.4}{0.7}$

22._____ $.4x = (0.08)(0.7)$ $X = 0.14$

23. $\dfrac{x}{3.8} = \dfrac{2.5}{5}$

23._____

24. $\dfrac{x}{14} = \dfrac{105}{138}$

24._____ $138x = (105)(14)$ $X = 10.65$

25. $\dfrac{0.9}{10.7} = \dfrac{4.3}{x}$

25._____

26. $\dfrac{175}{24.3} = \dfrac{x}{2.5}$

26._____ $24.3x = (175)(2.5)$ $X = 18.0$

Find the unknown number in each proportion. Write your answers as whole or mixed numbers when possible.

27. $\dfrac{16}{1\frac{1}{3}} = \dfrac{30}{x}$

27._____

28. $\dfrac{x}{\frac{7}{10}} = \dfrac{2\frac{2}{9}}{\frac{7}{3}}$ 20

$2\frac{1}{3}$

28._____

29. $\dfrac{3\frac{1}{3}}{1\frac{1}{2}} = \dfrac{x}{3\frac{1}{4}}$ 7   13

4

29._____

30. $\dfrac{1\frac{7}{8}}{x} = \dfrac{\frac{5}{12}}{\frac{5}{6}}$ 15

30._____

6.4   Problem Solving with Proportions

Section 6.4 Objectives
1. Use proportions to solve application problems.

Objective 1   Use proportions to solve application problems.

Set up and solve a proportion for each application problem.

1. If 3 ounces of a medicine must be mixed with 7 ounces of water, how many ounces of medicine would be mixed with 35 ounces of water?

1._____

2. The stock market report says that seven stocks went up for every five stocks that went down. If 825 stocks went down yesterday, how many went up?

2._____

$$\frac{7\,up}{5\,down} = \frac{x\,up}{825\,down} \qquad 5x = \frac{(7)(825)}{5}$$

$$x = 1155\ \text{went up}$$

3. The ratio of the length of an airplane wing to its width is 8 to 1. If the length of a wing is 37.5 meters, how wide must it be? Round to the nearest hundredth.

3._____

4. At 4 P.M., Yen's shadow is 1.45 meters long. Her height is 1.63 meters. At the same time, a tree's shadow is 7.12 meters long. How tall is the tree? Round to the nearest hundredth.

4._____

$$\frac{1.45\,(S)}{1.63\,(H)} = \frac{7.12\,(S)}{x\,(H)}$$

$$1.45x = (1.63)(7.12)$$

$$x = \frac{11.6}{1.45}$$

$$x = 8.0$$

5. Maxwell can sketch three portraits in seven hours. How long will it take him to sketch 10 portraits?

5._____

6. Seventy-five magazines cost $65. Find the cost of 12 magazines.

6.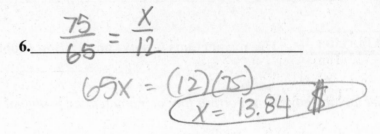

$$\frac{75}{65} = \frac{x}{12}$$

$$65x = (12)(75)$$

$$x = 13.84 \ \$$$

7. Craig makes $1035.93 in seven days. How much does he make in two days?

7._____

8. The Lincoln School District wants a student-to-teacher ratio of 14 to 1. How many teachers are needed for 2475 students? Round to the nearest whole number.

8.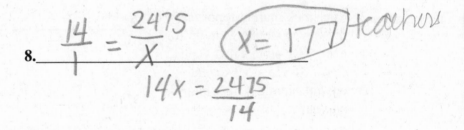

$$\frac{14}{1} = \frac{2475}{x} \qquad x = 177 \ \text{teachers}$$

$$14x = \frac{2475}{14}$$

9. Denise earns $1557.44 in 16 days. How much does she earn in 246 days?

9._____

10. The Lightning Lizards recorded seven songs on their first CD in 35 hours. How long will it take them to record 10 songs for their second CD?

10.

$$\frac{7}{35} = \frac{10}{x} =$$

$$7x = \frac{350}{7} \qquad x = 50 \ \text{hrs.}$$

11. If five pounds of fescue grass seed cover about 575 square feet of ground, how many pounds are needed for 5175 square feet?

11._____

12. Thirty-four flu vaccinations cost $425. Find the cost of 7 vaccinations.

12._____

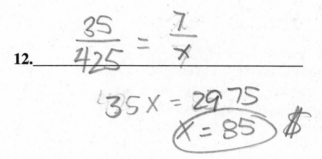

$$\frac{35}{425} = \frac{7}{x}$$

$$35x = 2975$$

$$x = 85 \quad \$$$

13. A survey showed that 3 out of 5 people would like to lose weight. At this rate, how many people in a group of 275 want to lose weight?

13._____

14. An advertisement says that 9 out of 10 dentists recommend sugarless gum. If the ad is true, how many of the 80 dentists in our city would recommend sugarless gum?

14._____

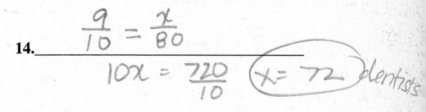

$$\frac{9}{10} = \frac{x}{80}$$

$$10x = \frac{720}{10} \qquad x = 72 \text{ dentists}$$

15. Cough syrup is to be given at the rate of 30 milliliters for each 100 pounds of body weight. How much should be given to a 56-pound child? Round to the nearest whole milliliter.

15._____

16. A U.S. map has a scale of 1 inch to 50 miles. Lake Birch is 1.75 inches long on the map. What is the lake's actual length in miles?

16. $\dfrac{1}{50} = \dfrac{1.75}{X}$

$50X = 87.5$ miles

17. In one state, 2 out of 7 college students receive financial aid. At this rate, how many of the 24,500 students at Redwood College receive financial aid?

17. _____

18. If 3 pounds of fertilizer will cover 75 square feet of garden, how many pounds are needed for 750 square feet?

18. $\dfrac{3}{75} = \dfrac{X}{750}$

$75X = 2250$

$\boxed{X = 30}$

6.5   Geometry: Lines and Angles

Section 6.5 Objectives
1. Identify and name lines, line segments, and rays.
2. Identify parallel and intersecting lines.
3. Identify and name angles.
4. Classify angles as right, acute, straight, or obtuse.
5. Identify perpendicular lines.
6. Identify complementary angles and supplementary angles and find the measure of a complement or supplement of a given angle.
7. Identify congruent and vertical angles and use this knowledge to find the measures of angles.
8. Identify corresponding angles and alternate interior angles and use this knowledge to find the measures of angles.

Key Terms *Answer the following questions about the key terms for Section 6.5.*

point   line   line segment   ray   parallel lines

intersecting lines   angle   degrees   straight angle

right angle   acute angle   obtuse angle   perpendicular lines

complementary angles   supplementary angles   congruent angles   vertical angles

corresponding angles   alternate interior angles

1. The most correct and complete difference   1._____
    between a **point** and a **line** is
    a. A **point** is a dot, but a **line** is a
        mark showing distance from A to
        B
    b. A **point** is the center of a circle,
        but a **line** is a side of a square
    c. A **point** is invisible, but a **line** is
        visible
    d. A **point** is a location in space, but
        a **line** is a continuous straight row
        of points in both directions

2. A line   2._____
    a. Goes from A to B
    b. Goes forever in both directions
    c. Always intersects another line
    d. Is another word for parallel

3. A **line segment**

    a. Has two endpoints

    b. Has one endpoint and goes forever in one direction

    c. Is written $\overrightarrow{AB}$

    d. Is one part of an angle

3._____

4. The correct way to write a **ray** is

    a. $\overleftarrow{AB}$

    b. angle AOB

    c. $\overline{AB}$

    d. $\overrightarrow{AB}$

4._____

5. Perpendicular lines differ from **intersecting lines** in what way?

    a. **Perpendicular lines** are always parallel

    b. **Intersecting lines** never cross each other

    c. **Perpendicular lines** always intersect to form a 90º angle

    d. **Perpendicular lines** always intersect to form an acute angle

5._____

6. An **angle** is formed when two _____ have a common endpoint.

    a. **Rays**

    b. **Lines**

    c. **Line segments**

    d. **triangles**

6._____

7. We measure the size of **angles** using what units?

    a. Inches

    b. Length

    c. Degrees

    d. Fractions

7._____

8. If I am driving north and turn my car 180º, what direction will I be driving?
    a. North
    b. East
    c. South
    d. West

8._____

9. Another term used to describe a 90º angle is
    a. **Straight** angle
    b. **Acute** angle
    c. **Obtuse** angle
    d. **Right** angle

9._____

10. An **angle** measuring 45º would be considered a/an
    a. **Straight** angle
    b. **Acute** angle
    c. **Obtuse** angle
    d. **Right** angle

10._____

11. An **angle** measuring 170º would be considered a/an
    a. **Straight** angle
    b. **Acute** angle
    c. **Obtuse** angle
    d. **Right** angle

11._____

12. Two streets that run **parallel** to each other would
    a. Never intersect
    b. Intersect at a 45º angle
    c. Intersect at a 90º angle
    d. Intersect at a curve

12._____

13. **Vertical angles** are always
    a. **Congruent**
    b. **Obtuse**
    c. **Transverse**
    d. **Corresponding**

13._____

14. In the figure, line *m* is parallel to line *n*.   **14.**_____
Angles 5 and 7 are

    a. **Supplementary angles**
    b. **Alternate interior angles**
    c. **Corresponding angles**
    d. **Vertical angles**

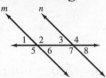

15. In the figure, line *m* is parallel to line *n*.   **15.**_____
Angles 6 and 3 are

    a. **Supplementary angles**
    b. **Alternate interior angles**
    c. **Corresponding angles**
    d. **Vertical angles**

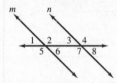

16. Two acute angles whose measures are   **16.**_____
added together can NOT be

    a. **Complementary angles**
    b. **Supplementary angles**
    c. **Straight angles**
    d. A **triangle**

17. Angles that measure the same number of   **17.**_____
degrees are called

    a. **Congruent angles**
    b. **Complementary angles**
    c. **Transverse angles**
    d. **Supplementary angles**

18. When angles are **supplementary**,   **18.**_____

    a. Both angles are obtuse
    b. The angles' measures add up to
        form a right angle
    c. Both angles are acute
    d. The angles' measures add up to
        form a straight angle

19. Vertical angles are formed by   19._____
intersecting lines. They are **congruent**
and
      a. Adjacent
      b. Supplementary
      c. Nonadjacent
      d. Complementary

Objective 1   **Identify and name lines, line segments, and rays.**

Identify each line, line segment, or ray and name it using the appropriate symbol.

1. (a)      1.(a)_____

1. (b)   A •————————• B   1.(b)_____

1. (c)   1.(c)_____

2. (a)      2.(a)_____

2. (b)   2.(b)_____

2. (c)   2.(c)_____

Objective 2   Identify parallel and intersecting lines.

Label each pair of lines as appearing to be parallel or intersecting.

3.

3._____

4.

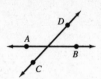

4._____

5.

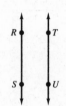

5._____

6.

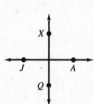

6._____

Objective 3   Identify and name angles.

Name each highlighted angle by using the three-letter form of identification.

7.

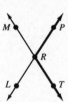

7._____

8.

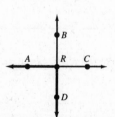

8._____

Name: Date:
Instructor: Section:

Objective 4   **Classify angles as right, acute, straight, or obtuse.**

Label each angle as acute, right, obtuse, or straight. For right angles and straight angles, indicate the number of degrees in the angle

9.(a)      **9.(a)**_____

9.(b)      **9.(b)**_____

9.(c)   **9.(c)**_____

10.(a)   **10.(a)**_____

10.(b)   **10.(b)**_____

10.(c)   **10.(c)**_____

Objective 5   **Identify perpendicular lines.**

In exercises 11 and 12, identify the pair of lines that is perpendicular. How can you describe the other pair of lines?

11.   **11.**_____

217

12.

12._____

Objective 6	Identify complementary angles and supplementary angles and find the measure of a complement or supplement of a given angle.

Identify each pair of complementary angles.

13.

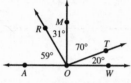

13._____

14.

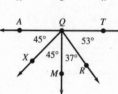

14._____

Identify each pair of supplementary angles.

15.

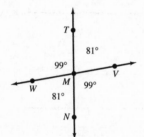

15._____

16.

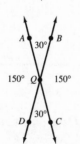

16._____

Find the complement of each angle.

    17. 38°   **17.**_____

    18. 79°   **18.**_____

    19. 11°   **19.**_____

    20. 50°   **20.**_____

Find the supplement of each angle.

    21. 12°   **21.**_____

    22. 91°   **22.**_____

    23. 160°   **23.**_____

    24. 1°   **24.**_____

| Objective 7 | Identify congruent angles and vertical angles and use this knowledge to find the measures of angles. |

In each figure, identify the angles that are congruent.

25.

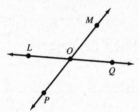

25._____

26.

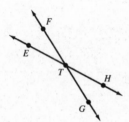

26._____

| Objective 8 | Identify corresponding angles and alternate interior angles and use this knowledge to find the measures of angles. |

In each figure, line m is parallel to line n. Find the measure of each angle.

27. $\angle 4$ measures $100°$.

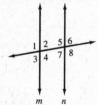

27._____

28. $\angle 7$ measures $37°$.

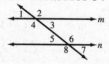

28._____

29. ∠6 measures 125°.

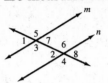

29. _____

30. ∠3 measures 44°.

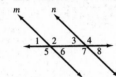

30. _____

6.6   Geometry Applications: Congruent and Similar Triangles

Section 6.6 Objectives

1. Identify corresponding parts of congruent triangles.
2. Prove that triangles are congruent using SAS, SSS, and ASA.
3. Identify corresponding parts of similar triangles.
4. Find the unknown lengths of sides in similar triangles.
5. Solve application problems involving similar triangles.

Key Terms

Answer the following questions about the key terms for Section 6.6.

congruent figures   **similar figures**

congruent triangles   **similar triangles**

1. **Congruent figures**   1._____
    a. Are the same shape and size
    b. Are the same shape but different
       sizes
    c. Are different shapes but the same
       size
    d. Have different perimeters but are
       the same shape

2. **Similar figures**   2._____
    a. Are the same shape and size
    b. Are the same shape but different
       sizes
    c. Are different shapes but the same
       size
    d. Have equal perimeters

3. The measurement of corresponding   3._____
   angles in **congruent triangles**
    a. Add up to 90°
    b. Are acute
    c. Are the same number of degrees
    d. Add up to 180°

4. In **similar triangles**, the *ratios* of the   4._____
   lengths of corresponding sides are
    a. Equal
    b. Different
    c. Not important
    d. Varying

223

Objective 1   Identify corresponding parts of congruent triangles.

Each pair of triangles is congruent. List the corresponding angles and the corresponding sides.

1.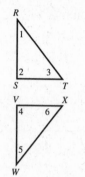

1._____

2.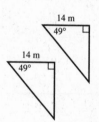

2._____

Objective 2   Prove that triangles are congruent using SAS, SSS, and ASA.

Determine which of these methods can be used to prove that each pair of triangles is congruent: Angle-Side-Angle (ASA), Side-Side-Side (SSS), or Side-Angle-Side (SAS).

3.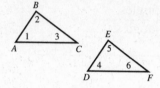

3._____

4.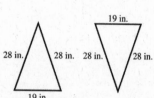

4._____

5.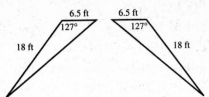

_____

6.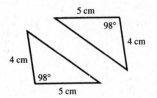

_____

7.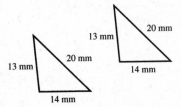

_____

8.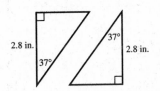

_____

9.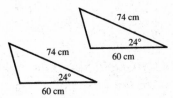

_____

10.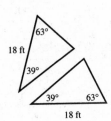

_____

Name: Date:

Instructor: Section:

Objective 3   Identify corresponding parts of similar triangles.

Write the ratio for each pair of corresponding sides in the similar triangles shown below. Write the ratios as fractions in lowest terms.

11. $\dfrac{ML}{AB}$; $\dfrac{MS}{AC}$; $\dfrac{LS}{BC}$ 11._____

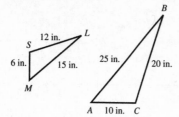

12. $\dfrac{BM}{FT}$; $\dfrac{BQ}{FA}$; $\dfrac{MQ}{TA}$ 12._____

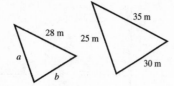

Objective 4   Find the unknown lengths of sides in similar triangles.

Find the unknown lengths in each pair of similar triangles.

13.   13._____

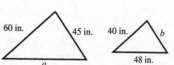

14.   14._____

226

15.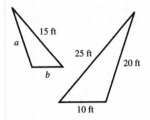

15._____

16.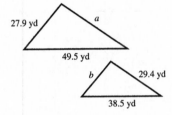

16._____

Find the perimeter of each triangle. Assume the triangles are similar.

17.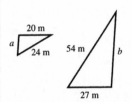

17._____

18.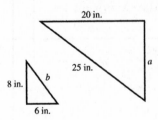

18._____

19.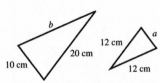

19._____

20.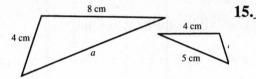

20._____

| Objective 5 | Solve application problems involving similar triangles. |

Solve each application problem.

21. A flagpole casts a shadow 52 m long at the same time that a pole 9 m tall casts a shadow 12 m long. Find the height of the flagpole.

21._____

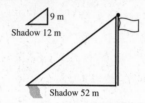

22. A flagpole casts a shadow 77 feet long at the same time that a pole 15 feet tall casts a shadow 55 ft long. Find the height of the flagpole.

22._____

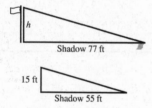

23. The height of the house shown here can be found by comparing its shadow to the shadow cast by a 5 ft stick. Find the height of the house by writing a proportion and solving it.

23._____

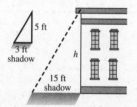

24. A fire lookout tower provides an
excellent view of the surrounding
countryside. The height of the tower
can be found by lining up the top of
the tower with the top of a
3-meter stick. Use similar triangles
to find the height of the tower.

24. _____

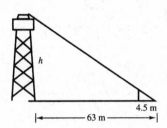

h

4.5 m

|← 63 m →|

Chapter 7 PERCENT

7.1   The Basics of Percent

Section 7.1 Objectives
1. Learn the meaning of percent.
2. Write percents as decimals.
3. Write decimals as percents.
4. Write percents as fractions.
5. Write fractions as percents.
6. Use 100% and 50%.

Key Terms: *Answer the following questions about the key terms for Section 7.1.*
              Percent
   1. Another way to say 25% is   1._____
       a. Twenty five out of one
           thousand
       b. A rate of twenty five miles
           per hour
       c. Twenty five out of one
           hundred
       d. One hundred twenty fives

Objective 1   **Learn the meaning of percent.**

Write a percent to describe each situation.

   1. You leave a $18 tip for a restaurant   1._____
       bill of $100. What percent tip did
       you leave?

   2. You earn 93 points on a 100-point   2._____
       test. What percent of the points did
       you earn?

Objective 2   **Write percents as decimals.**

Write each percent as a decimal.

3. (a) 15%

    (b) 8%

    (c) 260%

    (d) 0.3%

4. (a) 45%

    (b) 2%

    (c) 700%

    (d) 0.27%

3. (a)_____

   (b)_____

   (c)_____

   (d)_____

4. (a)_____

   (b)_____

   (c)_____

   (d)_____

Objective 3   **Write decimals as percents.**

Write each decimal as a percent.

5. (a) 0.7

    (b) 0.135

    (c) 4

    (d) 0.04

6. (a) 0.3

    (b) 0.936

    (c) 7.8

    (d) 0.05

5. (a)_____

   (b)_____

   (c)_____

   (d)_____

6. (a)_____

   (b)_____

   (c)_____

   (d)_____

Objective 4   **Write percents as fractions.**

Write each percent as a fraction or mixed number in lowest terms.

7. 60%

7._____

8. 25%

8._____

9. 75%

9._____

10. 62.5%

10._____

11. 12.5%

11._____

12. 55%

12._____

13. 275%

13._____

14. 350%

14._____

15. 5%

15._____

16. 8%

16._____

Objective 5   **Write fractions as percents.**

Write each fraction as a percent. If you're using a calculator, first work each one by hand. Then use your calculator and round to the nearest tenth of a percent if necessary.

17. $\dfrac{2}{5}$ 17._____

18. $\dfrac{1}{2}$ 18._____

19. $\dfrac{7}{10}$ 19._____

20. $\dfrac{3}{20}$ 20._____

21. $\dfrac{1}{25}$ 21._____

22. $\dfrac{3}{50}$ 22._____

23. $\dfrac{1}{9}$ 23._____

24. $\dfrac{6}{7}$ 24._____

Objective 6   Use 100% and 50%.

Fill in the blanks. Remember that 100% is all of something and 50% is half of it.

25   100% of $63 is _____.

25._____

26. 50% of 5 hours is _____.

26._____

27. 50% of 260 miles is _____.

27._____

28. 50% of $700 is _____.

28._____

29. 50% of 9 hours is _____.

29._____

30. 100% of 40 acres is _____.

30._____

7.2   The Percent Proportion

Section 7.2 Objectives

1. Identify the percent, whole, and part.
2. Solve percent problems using the percent proportion.

Key Terms: *Answer the following questions about the key terms for Section 7.2.*

percent proportion   whole   part

1. In the **percent proportion**   1._____

     a. the percent represents the **whole**

     b. the percent represents the **part**

     c. the **percent** is always 100

     d. the whole is the same as the

amount

2. The **part** and the **whole** in a **percent**   2. _____

     problem represent

       a. The **number** being compared
         to the **entire quantity**

       b. The **entire quantity** minus
         the **number**

       c. The **percent** plus the **entire
         quantity**

       d. The **base** divided by the
         amount

Objective 1   Identify the percent, whole, and part.

In Exercises 1-4, (a) identify the percent; (b) identify the whole; (c) identify the part.

1. Of the $3000, 12% will be   **1. (a) percent is** _____
     spent on a dining room table.

                                 (b) whole is _____

                                 (c) part is _____

2. $50 is 250% of what   **2. (a) percent is** _____
     number?

                                 (b) whole is _____

                                 (c) part is _____

3. 336 students is what percent of 840 students?

3. (a) percent is _____

(b) whole is _____

(c) part is _____

4. The state sales tax is $7\frac{3}{4}$ percent of the $895 price.

4. (a) percent is _____

(b) whole is _____

(c) part is _____

Objective 2   Solve percent problems using the percent proportion.

Write a percent proportion and solve it to answer these questions. If necessary, round money answers to the nearest cent and percent answers to the nearest tenth of a percent.

5. What is 20% of 4000 voters?

5. _____

6. 3% of 140 inches is how many inches?

6. _____

7. 19 falcons is what percent of 38 falcons?

7. _____

8. What percent of 365 days is 146 days?

8. _____

9. 357 fiction books is 42% of what number of books?

9. _____

10. $18\frac{1}{2}$% of what amount is $14.80?

10. _____

11. 350% of 10 days is how long?

11. _____

12. What percent of $250 is $64?

12. _____

13. What is 75% of 3260 athletes?

13. _____

14. 7% of $280 is how much money?

14. _____

15. 12 hours is what percent of 40 hours?

15. _____

16. What percent of 200 days is 32 cloudy days?

16. _____

17. 165% of what number of feet is 12.7 feet?

17. _____

18. What is 18.3% of $376?

18. _____

19. 207 employees is what percent of 59 employees?

19. _____

20. $0.28 is 9% of what number?

20. _____

21. 117 packages is 18% of what number of packages?

21. _____

22. $4\frac{1}{2}$% of what amount is $27.90?

22. _____

23. What is 120% of 80 acres?

23. _____

24. 7.3% of $47 is how much?

24. _____

25. What percent of 78 bicyclists is 110 bicyclists?

25. _____

26. 8% of what amount is $0.09?

26. _____

27. 1190 miles is 140% of what number of miles?

27. _____

28. $20 is what percent of $599?

28. _____

7.3   The Percent Equation

Section 7.3 Objectives
1. Estimate answers to percent problems involving 25%.
2. Find 10% and 1% of a number by moving the decimal point.
3. Solve basic percent problems using the percent equation.

Key Terms: *Answer the following questions about the key terms for Section 7.3.*

percent equation

1. The percent equation is shown   1._____
   correctly in which of the following?
   a. $20\% + \$5.00 = \$6.00$
   b. $\$5.00 - 20\% = \$4.00$
   c. $\$5.00 \div 20 = \$.25$
   d. $20\% \times \$5.00 = \$1.00$

Objective 1   **Estimate answers to percent problems involving 25%.**

Estimate the answer to each question.

1. What is 25% of $210.37?   1._____

2. 25% of 43 inches is how many   2._____
   inches?

3. Find 25% of 9.7 minutes.   3._____

4. What is 25% of $7823?   4._____

Objective 2   Find 10% and 1% of a number by moving the decimal point.

Find the exact answer to each question by moving the decimal point. Round money answers to the nearest cent if necessary.

5. What is 10% of $210.37?

5._____

6. 10% of 43 inches is how many inches?

6._____

7. Find 10% of 9.7 minutes.

7._____

8. What is 10% of $1823?

8._____

9. What is 1% of $210.37?

9._____

10. 1% of 43 inches is how many inches?

10._____

11. Find 1% of 9.7 minutes.

11._____

12. What is 1% of $1823?

12._____

Objective 3   Solve basic percent problems using the percent equation.

Write and solve an equation to answer each question.

13. 85% of 980 letters is how many letters?

13._____

14. 195 calls is what percent of 260 calls?

14._____

15. 570 students is 60% of how many students?

15._____

16. $15\frac{1}{2}\%$ of what number of people is 1240 people?

16._____

17. What is 82% of 1500 acres?

17._____

18. 3% of $720 is how much?

18._____

19. 21 combinations is what percent of 50 combinations?

19._____

20. What percent of $540 is $1323?

20._____

21. 24 magazines is 6% of what number of magazines?

21._____

22. 32% of 240 gallons is how many gallons?

22._____

23. $0.28 is what percent of $3.50?

23._____

24. How many customers
is 125% of
88 customers?

24. _____

25. 50% of what number of
televisions is
325 televisions?

25. _____

26. What percent of 600
kilometers is
11.1 kilometers?

26. _____

27. 120% of what number
of liters is
11.4 liters?

27. _____

28. What is 17.1% of 8500
acres?

28. _____

29. $8\frac{1}{2}$% of what number
of cartons is
51 cartons?

29. _____

30. 325% of what number of
liters is 26 liters?

30. _____

7.4   Problem Solving with Percent

Section 7.4 Objectives
1. Solve percent application problems.
2. Solve problems involving percent of increase or decrease.

Key Terms *Answer the following questions about the key terms for Section 7.4.*

percent of increase or decrease

1. The **percent of increase or**   1._____
   decrease is shown as a percent of
   a. The original value
   b. The amount of change
   c. The new value
   d. The reduced value

Objective 1   **Solve percent application problems.**

Use the six problem-solving steps to answer each question. Round percent answers to the nearest tenth of a percent.

1. 20% of the students at Lakeville   1._____
   Community College are married.
   How many of the 3260 students
   enrolled this year are married?

2. There were 75 points on the first   2._____
   statistics test. Steven's score was
   88% correct.
   How many points did Steven earn?

3. Lin needs 60 credits to graduate from   3._____
   her community college. So far she
   has earned 51 credits. What percent
   of the required credits does she have?

4. Total Fitness Club predicted that 240 new members would join after Christmas. It actually had 396 new members join. The actual number joining is what percent of the predicted number?

4._____

5. Jasmine did 40 problems correctly on a test, giving her a score of $62\frac{1}{2}\%$. How many problems were on the test?

5._____

6. Janel had budgeted $250 for new school clothes but ended up spending $390. The amount she spent was what percent of her budget?

6._____

7. David Clemens has 4.5% of his earnings deposited into a money market. If this amounts to $146.25 per month, find his monthly and annual earnings.

7._____

8. Louise Trent earns $450 per week and has 14% of this amount withheld for taxes. Find the amount withheld.

8._____

9. Members who are between 25 and 45 years of age make up 92% of the total membership of an organization. If there are 782 members in this group, find the total membership of the organization.

9._____

10. A survey at an intersection found that, of 2450 drivers, 12% were talking on a cell phone. How many drivers in the survey were talking on the phone?

10._____

Objective 2   **Solve problems involving percent of increase or decrease.**

Use the six problem-solving steps to find the percent increase or decrease. Round your answers to the nearest tenth of a percent.

11. A business increased the number of phone lines from 4 to 9. What is the percent increase?

11._____

12. Over the last three years, Calvin's salary has increased from $2700 per month to $3200. What is the percent increase?

12._____

13. During a sale, the price of a futon was cut from $1250 to $999. Find the percent of decrease in price.

13._____

14. Tomika's part-time work schedule has been reduced to 20 hours per week. She had been working 28 hours per week. What is the percent decrease?

14._____

15. Students at Withrow's College were charged $1560 for tuition this semester. If the tuition was $1480 last semester, find the percent of increase.

15._____

16. Tracy works as a massage therapist. During July, she cut her price on massages from \$54 to \$45.50. By what percent did she decrease the price?

16._____

17. During the holiday season, average daily attendance at a health club fell from 495 members to 230 members. What was the percent decrease?

17._____

18. Lewis has increased his exercise schedule from 5 hours per week to $10\frac{1}{2}$ hours per week. What is the percent increase?

18._____

7.5   Consumer Applications: Sales Tax, Tips, Discounts, and Simple Interest

Section 7.5 Objectives
1. Find sales tax and total cost.
2. Estimate and calculate restaurant tips.
3. Find the discount and sale price.
4. Calculate simple interest and the total amount due on a loan.

Key Terms *Answer the following questions about the key terms for Section 7.5.*

sales tax   tax rate   discount   interest   principal

interest rate   simple interest   interest formula

1. When calculating a 9% **sales tax** on a $6.00   1._____
    item,
    a.   Multiply 9 times 6; tax is $54.00
    b.   Multiply .9 times 6; tax is $5.40
    c.   Multiply .09 times 6; tax is $.54
    d.   Multiply .009 times 6; tax is $.05

2. The **tax rate** is expressed as a   2._____
    a. Percent
    b. Dollar amount
    c. Total cost
    d. Discount

3. If a $1500 television is on sale with a 15%   3._____
    discount, to calculate the new price,
    a. Multiply 1500 times 15 and divide by 2
    b. Multiply 1500 times 15 and subtract that
    amount from 1500
    c. Divide 1500 by 15 and subtract that amount
    from 1500
    d. Subtract 15% from 1500 and multiply by 5

4. If you pay 5% simple interest on a one-year $850   **4.**_____
    loan from your uncle, you will
    a. Subtract 5% from the total loan and pay that
    amount
    b. Multiply the loan by 5% to find the extra
    amount you will owe
    c. Multiply the loan by 5% and then by 12
    months to find the amount of interest you will
    add to the original loan
    d. Add $85 to the original amount to find your
    total amount due

5. An **interest rate** is usually given as a percent and   **5.**_____
    a. assumed to be **per year**
    b. assumed to be **per month**
    c. assumed to be **per week**
    d. assumed to be **per day**

6. The **interest formula** is   **6.**_____
    a. $I = p \cdot r$
    b. $I = p \cdot r \div t$
    c. $I = \dfrac{p}{r} \cdot t$
    d. $I = p \cdot r \cdot t$

Objective 1   Find sales tax and total cost.

Find the amount of the sales tax or the tax rate and the total cost. Round money answers to the nearest cent.

	Cost of Item	Tax Rate	Amount of Tax	Total Cost
1.	$100	8%	_____	_____
2.	$300	5%	_____	_____
3.	$98	_____	$1.92	_____

4. $175   _____ $12.25   _____

5. $32.49   6%   _____   _____

6. $3.25 $4\frac{1}{2}\%$ _____   _____

7. $14,200   _____ $923   _____

8. $23,500   _____ $822.50   _____

Objective 2   **Estimate and calculate restaurant tips.**

For each restaurant bill, estimate a 15% tip and a 20% tip. Then find the exact amounts for a 15% tip and a 20% tip. Round exact amounts to the nearest cent if necessary

	Bill	Estimate of 15% tip	Exact 15% tip	Estimate of 20% tip	Exact 20% tip
9.	$43.16	_____	_____	_____	_____
10.	$87.22	_____	_____	_____	_____
11.	$63.85	_____	_____	_____	_____
12.	$72.81	_____	_____	_____	_____
13.	$23.95	_____	_____	_____	_____
14.	$9.62	_____	_____	_____	_____

Objective 3   Find the discount and sale price.

Find the amount or rate of discount and the sale price. Round money answers to the nearest cent if necessary.

	Original Price	Rate of Discount	Amount of Discount	Sale Price
15.	$100	25%	_____	_____
16.	$260	30%	_____	_____
17.	$49.99	10%	_____	_____
18.	$28.43	15%	_____	_____
19.	$15.50	_____	$1.86	_____
20.	$99	_____	$59.40	_____
21.	$32	_____	$4.80	_____
22.	$250	_____	$100	_____

Objective 4   Calculate simple interest and the total amount due on a loan.

Find the simple interest and total amount due on each loan.

	Principal	Rate	Time	Interest	Total Amount Due
23.	$300	12%	1 year	_____	_____
24.	$800	8%	2 years	_____	_____
25.	$1260	7%	6 months	_____	_____
26.	$940	6%	9 months	_____	_____

	Principal	Rate	Time	Interest	Total Amount Due
27.	$18,500	$7\frac{1}{2}\%$	3 years	_____	_____
28.	$24,800	$6\frac{1}{2}\%$	4 years	_____	_____
29.	$8300	12%	6 months	_____	_____
30.	$650	10%	9 months	_____	_____

Chapter 8 MEASUREMENT

8.1   Problem Solving with English Measurement

Section 8.1 Objectives
1. Learn the basic measurement units in the English system.
2. Convert among measurement units using multiplication or division.
3. Convert among measurement units using multiplication or division.
4. Solve application problems using English measurement.

Key Terms: *Answer the following questions about the key terms for Section 8.1.*

English system   metric system   unit fraction

1. Examples of **English system of**   **1.**_____
measurement units are
        a. liters, grams, inches
        b. centimeters, kilograms,
        quarts
        c. gallons, inches, ounces
        d. yards, pounds, meters

2. The **metric system** is based on   **2.**_____
        a. multiples of 5
        b. multiples of 10
        c. multiples of 100
        d. multiples of 1000

3. A **unit fraction**   **3.**_____
        a. has a denominator of 1
        b. is equivalent to 1
        c. has the number you want to
change in the numerator
        d. should have the same units
in the numerator and in the
denominator

Objective 1   Learn the basic measurement units in the English system.

Fill in the blanks with the measurement relationships you have memorized.

    1.(a) 1 yd = _____ ft   **1. (a)**_____

    (b)   1 hr = _____ min   **(b)**_____

    (c)   _____ c = 1 pt   **(c)**_____

2.(a) 1 lb = _____ oz   **2. (a)**_____

   (b) 1 gal = _____ qt   **(b)**_____

   (c) _____ days = 1 wk   **(c)**_____

3.(a) _____ fl oz = 1 c   **3. (a)**_____

   (b) _____ in. = 1 ft   **(b)**_____

   (c) 1 mi = _____ ft   **(c)**_____

4.(a) _____ pt = 1 qt   **4. (a)**_____

   (b) 1 day = _____ hr   **(b)**_____

   (c) _____ sec = 1 min   **(c)**_____

Objective 2   **Convert among measurement units using multiplication or division.**

Convert each measurement using multiplication or division.

   5. 3 T to pounds   **5.**_____

   6. 15 pt to quarts   **6.**_____

   7. 32 oz to pounds   **7.**_____

   8. 40 min to hours   **8.**_____

   9. $12\frac{1}{2}$ ft to inches   **9.**_____

   10. $4\frac{1}{2}$ lb to ounces   **10.**_____

Objective 3   **Convert among measurement units using unit fractions.**

Convert each measurement using unit fractions.

11. 4 c to pints   11._____

12. 20 qt to gallons   12._____

13. $5\frac{1}{2}$T to pounds   13._____

14. 8 oz to pounds   14._____

15. 120 min to hours   15._____

16. $2\frac{3}{4}$lb to ounces   16._____

17. 14 qt to gallons   17._____

18. 180 sec to minutes   18._____

19. 9 in. to feet   19._____

20. 15 yd to feet   20._____

Use two or three unit fractions to convert the following.

21. 5 days = _____ sec   **21.**_____

22 2 mi = _____ in.   **22.**_____

23. 420 hr = _____ wk   **23.**_____

24. 20 pt = _____ gal   **24.**_____

Objective 4   Solve application problems using English measurement.

Solve each application problem. Show your work.

25. A 42 oz jar of peanuts is on sale for   **25.**_____
$6.79. What is the cost per pound, to
the nearest cent?

26. Tonette is making six matching table   **26.** _____
runners. If each runner requires 8 feet
of material and the material sells for
$4.99 per yard, how much will she
spend to make the runners?

27. Calvin is making four batches of a   **27.** _____
potato dish recipe for a holiday meal.
If each recipe requires $2\frac{3}{4}$ cups of
milk, how many quarts of milk will he
need?

28. Last year a city had 84 in. of snowfall while this year the city had 51 in. What is the difference in snowfall between the two years, in feet?

28. _____

29. A 20 oz package of chocolate chips is on sale for $4.89. What is the cost per pound, to the nearest cent?

29. _____

30. Adrian needs 5 feet of wire fencing to support each of 10 small trees. If the fencing sells for $10.39 per yard, how much will it cost to install fencing around all the trees, to the nearest cent?

30. _____

8.2   The Metric System—Length

Section 8.2 Objectives
1. Learn the basic metric units of length.
2. Use unit fractions to convert among units.
3. Move the decimal point to convert among units.

Key Terms: *Answer the following questions about the key terms for Section 8.2.*

meter	prefixes	metric conversion line

1. A **meter** is a measure of   1._____
     a. weight
     b. volume
     c. temperature
     d. length

2. Choose all of the correctly written   2 _____
abbreviations below
     a. M
     b. mm
     c. cm
     d. Km
     e. mtr
     f. dam

3. A **decimeter** is   3._____
     a. one full meter
     b. one-tenth of a meter
     c. ten meters
     d. one-hundredth of a meter

4. Using the **metric conversion line**   4._____
from your textbook, which statement is
true?
     a. a km is larger than a m
     b. a m is the smallest unit
     c. a dm is equal to a dam
     d. none of the above

Objective 1   Learn the basic metric units of length.

Write the most reasonable metric length unit in each blank. Choose from km, m, cm, and mm

1. The bookshelf is 2 _____ tall.

1._____

2. Kelly walks 5 _____ each day.

2._____

3. The Carpenter family drove 650 _____ on vacation.

3._____

4. The adult is 180 _____ tall.

4._____

5. The paper clip is 30 _____ long.

5._____

6. The pencil is 15 _____ long.

6._____

7. The hallway is 2 _____ wide.

7._____

8. The slice of cheese is 4 _____ thick.

8._____

9. The width of the pen is 8 _____

9._____

10. The broom handle is 1
_____ long.

10._____

Objective 2   **Use unit fractions to convert among units.**

Convert each measurement. Use unit fractions.

11. 5 m to mm

11._____

12. 300 mm to cm

12._____

13. 8.6 km to m

13._____

14. 25 cm to m

14._____

15. 50 mm to m

15._____

16. 0.2 km to m

16._____

17. 0.83 m to mm

17._____

18. 21 m to cm

18._____

19. 9 mm to m

19._____

20. 6 m to cm

20._____

Objective 3 Move the decimal point to convert among units.

Convert each measurement. Use the metric conversion line.

21. 60 m to km   21._____

22. 18.3 km to m   22._____

23. 5 m to mm   23._____

24. 40.8 cm to m   24._____

25. 3.99 m to cm   25._____

26. 11.002 km to m   26._____

27. 38 mm to cm   27._____

28. 804 cm to m   28._____

29. 7 cm to m   29._____

30. 0.6 m to cm   30._____

8.3   The Metric System—Capacity and Weight (Mass)

Section 8.3 Objectives
1. Learn the basic metric units of capacity.
2. Convert among metric capacity units.
3. Learn the basic metric units of weight (mass).
4. Convert among metric weight (mass) units.
5. Distinguish among basic metric units of length, capacity, and weight (mass).

Key Terms: *Answer the following questions about the key terms for Section 8.3.*

                  liter   gram

1. A **liter** is to capacity as   1._____
    a. a meter is to length
    b. a centimeter is to degrees
    c. a kilogram is to volume
    d. a degree Celsius is to capacity

2. An object likely to weigh about one   2._____
gram is
    a. a loaf of bread
    b. a newborn baby
    c. a permanent marker
    d. a dollar bill

| Objective 1 | **Learn the basic metric units of capacity.**

Write the most reasonable metric length unit in each blank. Choose from L and mL.

1.(a) The bathtub holds 80   **1.(a)**_____
    _____ of water.
                                                    (b)_____
 (b) She drank 200 _____ of
    juice.

2.(a) The door prize was a 2
_____ bottle of soda.

(b) Ben gave his daughter 15
_____ of cough syrup.

2.(a)_____

(b)_____

3.(a) The dog drank 500
_____ of water.

(b) I bought 12 _____ of
stain for the patio.

3.(a)_____

(b)_____

4.(a) The recipe called for 15
_____ of lemon
extract.

(b) We'll need 20 _____
of soda for the picnic.

4.(a)_____

(b)_____

Objective 2   **Convert among metric capacity units.**

Convert each measurement. Use unit fractions or the metric conversion line.

   5. 300 mL to L

   6. 12.8 L to mL

   7. 37.15 L to mL

   8. 834 mL to L

   9. 16 L to mL

5._____

6._____

7._____

8._____

9._____

10. 4000 mL to L   **10.**_____

11. 76,000 mL to L   **11.**_____

12. 2 L to mL   **12.**_____

Objective 3   **Learn the basic metric units of weight (mass).**

Write the most reasonable metric unit in each blank. Choose from kg, g, and mg.

13.(a) Two apples weigh 500   **13.(a)**_____
_____ .

(b). The staple weighs 300   **(b)**_____
_____.

                                         (c)_____

(c) The professional baseball player
weighs 80 _____.

14.(a) The nail weighs 3   **14.(a)**_____
_____ .

(b) The football weighs   **(b)**_____
500_____.

                                         (c)_____

(c) The airline has a limit of 25
_____ per piece of
luggage.

Objective 4   Convert among metric weight (mass) units.

Convert each measurement. Use unit fractions or the metric conversion line.

15. 18.32 kg to g   15._____

16. 4.1 kg to g   16._____

17. 0.3 g to mg   17._____

18. 3400 mg to g   18._____

19. 9000 g to kg   19._____

20. 20,000 mg to g   20._____

21. 0.093 g to mg   21._____

22. 30 g to kg   22._____

Objective 5   Distinguish among basic metric units of length, capacity, and weight (mass).

Write the most appropriate metric unit in each blank. Choose from km, m, cm, mm, L, mL, kg, g, and mg.

23. The belt is 1   23._____
    _____ long.

24. A quarter is about 24
_____ across.

24._____

25. The plane flew 700
_____.

25._____

26. Add 10 _____ of
soy sauce to the stir fry.

26._____

27. Her glasses weigh 25
_____.

27._____

28. Buy 3 _____ of
milk at the store.

28._____

29. The dollar bill is about 6.5
_____ wide.

29._____

30. Each television weighs 7
_____ .

30._____

8.4   Problem Solving with Metric Measurement

Section 8.4 Objectives
1. Solve application problems involving metric measurements.

Objective 1   **Solve application problems involving metric measurements.**

Solve each application problem. Show your work. Round money answers to the nearest cent.

1. Sliced turkey meat is on sale at
    $9.89 per kilogram. Nelson bought
    750 g of the turkey. How much did
    he pay, to the nearest cent?

1._____

2. Francis has 3.8 m of braided
    cording. How many centimeters of
    cording can she use to trim each of
    four napkins assuming she uses the
    same amount on each napkin?

2._____

3. Rosalee is building a table. She has
    one board that is 114 cm 14 cm and
    another that is 289 cm. How long
    are the two boards together, in
    meters?

3._____

4. Mary Beth needs 270 cm of
    material to make a skirt. The price
    is $6.49 per meter plus a 6% sales
    tax. How much will she pay, to the
    nearest cent?

4._____

5. Jannella's doctor wants her to take
    2.8 g of medication each day in
    four equal doses. How many
    milligrams should be in each dose?

5._____

6. Bulk beans are on sale at $0.85 per kilogram. Rochelle scooped some beans into a bag and put it on the scale. How much will she pay for 3 kg 20 g of beans, to the nearest cent?

6._____

7. Brian has two pieces of rope. One measures 3 m 35 cm and the other measures 4 m 72 cm. How many meters of rope does he have in all?

7._____

8. Which case of pasta sauce is the better buy: a $29.50 case that holds ten 1 L bottles or a $28 case that holds sixteen 600 mL bottles? What is the price per liter?

8._____

9. Luis is catering a lunch for 25 people. He plans to have 250 mL of lemonade punch for each guest. How many liters of punch should he make?

9._____

10. Ahmed drinks two 8 oz cups of double espresso each morning on the way to work. Each cup contains 160 mg of caffeine, and he works six days per week. How many grams of caffeine does he get from the espresso each week?

10._____

11. Coretta is studying mosquitoes that can transmit diseases. One type of mosquito averages 1.5 cm in length. Another kind averages only 0.9 cm in length. What is the difference, in millimeters, between the lengths of the mosquitoes?

11._____

12. Linda bought 5 m of garden fencing. She used 3 m of the fencing around some rose bushes and 80 cm around the trunk of a tree. How much fencing is left, in meters?

12._____

13. A rectangular table in the biology lab measures 120 cm long by 84 cm wide. How much will it cost to put a rubber strip around all four edges of the table? The rubber strip costs $8.98 per meter plus 7% sales tax.

13._____

14. A stew recipe makes 12 servings and calls for 2 kg of beef. How much meat is used for each serving, to the nearest gram?

14._____

15. A man's healthy heart, at rest, is beating 60 times per minute. If his heart pumps about 4.2 L of blood each minute, how many milliliters are pumped per beat?

15._____

16. Yao usually drinks three cans of soda and two large glasses of milk each day. A soda can holds 350 mL and a large glass holds 425 mL. How many liters does he drink n one day? In one week?

16._____

8.5   Metric–English Conversions and Temperature

Section 8.5 Objectives
1.   Use unit fractions to convert between metric and English units.
2.   Learn common temperatures on the Celsius scale.
3.   Use formulas to convert between temperatures.

Key Terms: *Answer the following questions about the key terms for Section 8.5.*

Celsius	Fahrenheit

1. On a Fahrenheit scale, freezing is   1._____
      a. ⁻10°F
      b. 32°F
      c. 0°F
      d. ⁻32°F

2. On a Celsius scale, a comfortable   2._____
outdoor temperature would be
      a. 100°C
      b. 70°C
      c. 50°C
      d. 30°C

Objective 1   **Use unit fractions to convert between metric and English units.**

Use the tables on the first page of Section 8.5 in the textbook and unit fractions to make approximate conversions from metric to English or English to metric. Round your answers to the nearest tenth.

   1. 18 ft to meters   1._____

   2. 4.5 yd to meters   2._____

   3. 12.25 L to gallons   3._____

   4. 38.2 L to quarts   4._____

5. 30 m to yards.   5._____

6. 260 g to ounces   6._____

7. 3.21 kg to pounds   7._____

8. 60 m to feet   8._____

9. 4.5 oz to grams   9._____

10. 9 km to miles   10._____

11. 20 cm to inches   11._____

12. 8 mi to kilometers   12._____

| Objective 2 | **Learn common temperatures on the Celsius scale.**

Circle the most reasonable Celsius temperature for each situation.

13. Hot cocoa
    65°C   65°F   110°C 200°C

14. Normal body temperature
    20° C   37°C   98° C

15. Nice spring day
    0°C   14°C   50°C

16. Heavy coat weather
    ⁻10°C   5°C   22°C

17. Boiling water for tea
    32°C   75°C   100°C

18. Hot tub water
    5°C   30°C   55°C

19. Inside a freezer
    ⁻8°C   4°C   25°C

20. Weather for wearing shorts and
    sandals
    10°C   33°C   90°C

Objective 3   **Use formulas to convert between temperatures.**

Use the conversion formulas and the order of operations to convert Fahrenheit temperatures to Celsius and Celsius temperatures to Fahrenheit. Round your answers to the nearest degree if necessary.

21. 30°F   21._____

22. 85°C   22._____

23. 82°F   23._____

24. 9°C   24._____

25. 25°F   25._____

26. –21°C   26._____

27. 12°C   27._____

28. 0°F   28._____

29. -7°C   29._____

30. 15°F   30._____

Chapter 9   GRAPHS

9.1   Problem Solving with Tables and Pictographs

Section 9.1 Objectives
1. Read and interpret data presented in a table.
2. Read and interpret data from a pictograph.

Key Terms: *Answer the following questions about the key terms for Section 9.1.*

table   pictograph

1. When reading a **table**   1._____

    a. read from bottom to top

    b. read from left to right, and check the column headings

    c. read only the right-hand column

    d. it doesn't matter what direction you read it

2. A **pictograph**   2._____

    a. easily shows fractional amounts of a symbol

    b. uses rounding in order to be more exact

    c. should be used for general comparisons and approximations

    d. is the most accurate type of graph

Objective 1   **Read and interpret data presented in a table.**

The table below shows information about the performance of the seven largest U.S. airlines during the second quarter of 2004 (April–June). Use this table to answer these questions.

PERFORMANCE DATA FOR THE LARGEST U.S. AIRLINES APRIL–JUNE 2004		
Airline	On-Time Performance	Luggage Handling*
American	72%	4.3
Continental	77%	3.0
Delta	78%	3.1
Northwest	78%	3.7
Trans World	74%	4.0
United	57%	5.3
US Airways	71%	4.1

*Luggage problems per 1000 passengers

1. What percent of United flights were on
   time?

1._____

2. What percent of Northwest flights were
   on time?

2._____

3. Which airline(s) had more than 5 luggage
   handling problems per 1000 passengers?

3._____

4. Which airline had the worst on-time
   performance?

4._____

5. What is Delta's luggage handling record?

5._____

6. Which airline(s) had on-time performance
   of at least 75%?

6._____

7. Which airline(s) had on-time performance
   of at most 70%?

7._____

8. What was the average on-time
   performance for all seven airlines, to the
   nearest tenth of a percent?

8._____

This table show the maximum cab fares in five different cities in February 2004. The "flag drop" charge is made when the driver starts the meter. "Wait time" is the charge for having to wait in the middle of a ride. Use the table to answer these questions.

MAXICAB FARES ALLOWED IN SELECTED CITIES IN FEBRUARY 2004			
City	Flag Drop	Price per Mile	Wait Time per Hour
Chicago	$1.90	$1.60	$20
Denver	$1.60	$1.80	$18
Miami	$1.70	$2.20	$21
New York	$2.50	$2.00	$12
Portland	$2.50	$1.50	$20

9. What is the maximum fare for an 8-mile ride in Miami that includes 20 minutes of wait time?

9._____

10. What is the maximum fare for a 5.5-mile cab ride with 15 minutes of wait time) in Chicago?

10._____

11. Find the cost of a cab ride of 7 miles in Denver, plus a 15% tip.

11._____

12. What is the difference in the maximum fare for a cab ride of 3 miles in Portland compared to New York? Assume there is no wait time.

12._____

Objective 2   Read and interpret data from a pictograph.

The pictograph below shows the population of five cities in 2003. Use the pictograph to answer these questions.

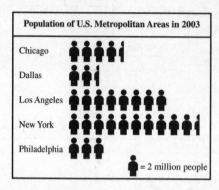

13. What is the approximate population of Los Angeles?

13._____

14. What is the approximate population of Dallas?

14._____

15. Approximately how much greater is the population of Los Angeles than Chicago?

15._____

16. Approximately how much greater is the population of New York than Dallas?

16._____

17. The population of Philadelphia is approximately how much less than Chicago?

17._____

18. What is the approximate total population of all five cities?

18._____

This pictograph shows the approximate number of passenger arrivals and departures at selected U.S. airports in 2004. Use the pictograph to answer these questions.

**Passenger Arrivals and Departures at
Selected U.S. Airports in 2004**

Atlanta	▭ ▭ ▭ ▭ ▭ ▭ ▭ ⊏
Dallas/Ft. Worth	▭ ▭ ▭ ▭ ▭ ▭
Miami	▭ ▭ ▭ ⊏
San Francisco	▭ ▭ ▭ ▭
St. Louis	▭ ▭ ▭

▭ = 10 million passenger
arrivals and departures

19. Approximately how many passenger arrivals and departures took place at the San Francisco airport?

19._____

20. What is the approximate total number of arrivals and departures at the two least busy airports?

20._____

21. What is the difference in the number of arrivals and departures at Dallas/Ft. Worth airport and Miami airport?

21._____

22. Find the average number of arrivals and departures for the three busiest airports.

22._____

9.2   Reading and Constructing Circle Graphs

Section 9.2 Objectives
1. Read a circle graph.
2. Use a circle graph.
3. Use a protractor to draw a circle graph.

Key Terms: *Answer the following questions about the key terms for Section 9.2.*

circle graph   protractor

1. A **circle graph** is divided into   1._____
   a. sectors
   b. sections
   c. sects
   d. percents

2. When using a **protractor**   2._____
   a. never move it
   b. always make sure your marks add up to 180°
   c. always make sure your marks add up to 90°
   d. be sure to put the hole in the exact center of the circle

Objective 1   **Read a circle graph.**

The following circle graph shows you how 30 children in a day care center are divided among different age groups. Use this circle graph to answer these questions.

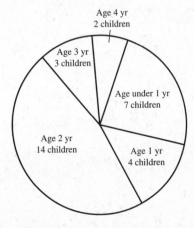

1. The greatest number of children are in which
   age group?   1._____

2. The fewest number of children are in which
   age group?   2._____

3. How many more children are in the 1-year-
   old age group than in the 4-year-old age
   group?   3._____

4. Find the total number of children that are
   less than 3 years old.   4._____

Objective 2   **Use a circle graph.**

The following circle graph shows you how 30 children in a day care center are divided among different age groups. Use this circle graph to find each ratio. Write ratios as fractions in lowest terms.

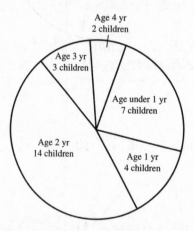

Age 4 yr
2 children

Age 3 yr
3 children

Age under 1 yr
7 children

Age 2 yr
14 children

Age 1 yr
4 children

5. Number of 4-year-olds to total
   number in daycare.   5._____

6. Number of 2-year-olds to total   **6.**_____
   number in daycare.

7. Total number of children in daycare **7.**_____
   to number of children older than 2
   years.

8. Number of children 2 years old to   **8.**_____
   number of children 4 years old.

The following circle graph shows you how a monthly income for a family is divided among different categories. Each cost is expressed as a percent of the total of $4000. Use this circle graph to answer these questions.

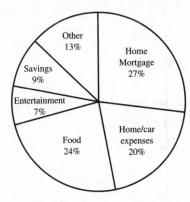

 9. How much is spent each month on   **9.**_____
    food?

10. How much is spent each month on   **10.**_____
    entertainment?

11. What is the largest single expense? **11.**_____

12. What is the smallest expense?

12._____

13. What is the ratio of the mortgage, home and car expenses to the total monthly income?

13._____

14. What is the ratio of food to savings and entertainment?

14._____

The following circle graph shows you how customers in a survey said they heard about a certain store. If 4350 people were surveyed, use the graph to find the number in each of the following categories. Round to the nearest whole number.

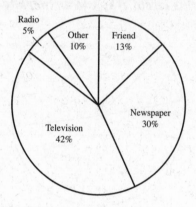

15. Friend

15._____

16. Newspaper

16._____

17. Television

17._____

18. Radio

18._____

19. Newspaper and Television   19._____
    combined

20. Name the highest category and   20._____
    the lowest category. How many
    more people are in the highest
    category than the lowest?

| Objective 3 | Use a protractor to draw a circle graph. |

21. *During a one-year period, Quality Fixtures Inc. had the following expenses.*
Find all the missing numbers from the table.

	Item	Dollar Amount	Percent of Total	Degrees of a Circle
(a)	Salaries	$224,000	35%	_____
(b)	Delivery expense	$96,000	15%	_____
(c)	Advertising	$64,000	10%	_____
(d)	Rent	$160,000	_____	90°
(e)	Other	$96,000	_____	54°
(f)	Draw a circle graph using a protractor and the information above.			

22. During one semester Yolanda Cook spent $6200 for school expenses as shown in this table. Find all the missing numbers from the table.

	Item	Dollar Amount	Percent of Total	Degrees of a Circle
(a)	Rent	$1240	20%	_____
(b)	Food	$1550	_____	90°
(c)	Clothing	$310	_____	_____
(d)	Books	$930	15%	_____
(e)	Entertainment	$310	5%	_____
(f)	Savings	$620	_____	_____
(g)	Tuition	$1240	_____	72°
(h)	Draw a circle graph using a protractor and information above.			

9.3   Bar Graphs and Line Graphs

Section 9.3 Objectives
1. Read and understand a bar graph.
2. Read and understand a double-bar graph.
3. Read and understand a line graph.
4. Read and understand a comparison line graph.

Key Terms: *Answer the following questions about the key terms for Section 9.3.*

> **bar graph**   **double-bar graph**   **line graph**
>
> **comparison line graph**

1. A **bar graph** shows bars of   1._____
different heights to represent
      a. quantity or frequency
      b. quantity and quality

2. A good type of graph to show the   2._____
sales of a shoe store for each month
over two years is
      a. a pictograph
      b. a circle graph
      c. a bar graph
      d. a double-bar graph

3. A graph showing how two or more   3._____
sets of data relate or compare to each
other is called a
      a. **pictograph**
      b. **circle graph**
      c. **bar graph**
      d. **comparison line graph**

4. Why is the label on the left side of   4._____
a **line graph** important to note?

Name: Date:
Instructor: Section:

Objective 1   Read and understand a bar graph.

This bar graph shows the results of a survey of 3000 people questioned on their favorite activities during time off from their jobs. Use the graph to answer these questions.

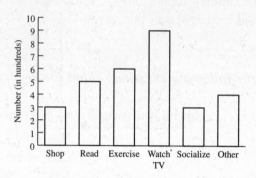

1. How many people enjoy socializing during their time off?

    1._____

2. How many people enjoy reading during their time off?

    2._____

3. What is the most popular activity during time off? How many chose this activity?

    3._____

4. What is the second most popular activity during time off? How many chose this activity?

    4._____

5. What percent of the people enjoy shopping during their time off?

    5._____

6. What percent of the people chose watching TV?

    6._____

Objective 2   **Read and understand a double-bar graph.**

This double-bar graph shows the number of workers that were unemployed in a city during the first six months of 2004 and 2005. Use this graph to answer these questions.

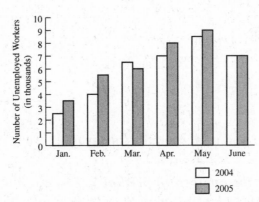

7. In which month in 2004 were the most workers unemployed? What was the total number unemployed in that month?

7._____

8. How many workers were unemployed in April of 2005?

8._____

9. How many more workers were unemployed in January of 2005 than in January of 2004?

9._____

10. How many fewer workers were unemployed in March of 2005 than in March of 2004?

10._____

11. Find the amount of increase and the percent of increase in the number of unemployed workers from February 2004 to May of 2004.

11._____

12. Find the amount of increase and
the percent of increase in the
number of unemployed workers
from January of 2005 to June of
2005.

12._____

Objective 3   **Read and understand a line graph.**

This line graph shows the number of speeding tickets given in a city during a
five-month period. Use the line graph to answer these questions. Round percent
answers to the nearest whole percent.

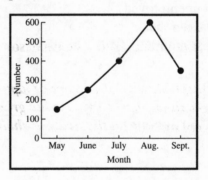

13. In what month were the most
tickets issued? How many
tickets were issued that
month?

13._____

14. In what month were the
fewest tickets issued? How
many tickets were issued that
month?

14._____

15. How many tickets were issued
in May and June combined?

15._____

16. How many more tickets were issued in July than in June?

16._____

17 Find the amount of decrease and the percent of decrease in the number of tickets from August to September.

17._____

18. Find the amount of increase and the percent of increase in the number of tickets from July to August.

18._____

| Objective 4 | **Read and understand a comparison line graph.**

This comparison line graph shows the number of laptop computers sold by two different stores during each of four years. Use the graph to answer these questions. Round percent answers to the nearest whole percent..

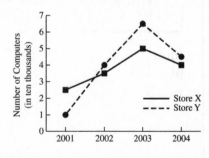

19. Sales at Store X in 2002

19._____

20. Sales at Store Y in 2004

20._____

21. Sales at Store Y in 2001

21._____

22 Sales at Store X in 2003

22._____

23. Amount and percent of increase in sales at Store Y from 2002 to 2003

23._____

24. Amount and percent of decrease in sales at Store X from 2003 to 2004?

24._____

9.4   The Rectangular Coordinate System

Section 9.4 Objectives
1. Plot a point, given the coordinates, and find the coordinates, given a point.
2. Identify the four quadrants and determine which points lie within each one.

Key Terms: *Answer the following questions about the key terms for Section 9.4.*

paired data	**horizontal axis**	**vertical axis**
x-axis **y*-axis**	**coordinate system**	**origin**
coordinates **quadrants**		

1. When writing **paired data**, write the pair   1._____
      a. using only words
      b. listing the vertical axis first, then the horizontal
      axis
      c. inside parentheses, separated by a comma
      d. using positive numbers

2. The **horizontal axis**   2._____
      a. goes up and down
      b. goes left and right

3. The **vertical axis**   3._____
      a. goes up and down
      b. goes left and right

4. When finding the location of an ordered pair   4._____
      a. the first number indicates the position on the
      vertical axis
      b. the second number indicates the position on the
      vertical axis
      c. the numbers may be interchanged with no
      effect on the location of the point
      d. the first number is always negative

Name: Date:
Instructor: Section:

Match the following.   **(a)** a rectangular
_____**5**. *x*-axis   coordinate system
_____**6**. *y*-axis   **(b)** vertical axis
_____**7**. origin   **(c)** 0.0
_____**8**. the *x*-axis and the *y*-axis together   **(d)** horizontal axis
_____**9**. the divisions of the coordinate system   **(e)** quadrants

| Objective 1 | **Plot a point, given the coordinates, and find the coordinates, given a point.** |

Plot each point on the rectangular coordinate system. Label each point with its coordinates.

1. (1, 3)

2. (–2, 0)

3. (–5, 4)

4. (0, 6)

5. (2, 5)

6. (4, –1)

7. (–5, –5)

8. (1, –6)

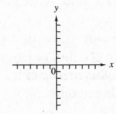

Give the coordinates of each point.

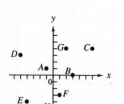

9. A_____

10. B_____

11. C_____

12. D_____

13. E_____

14. F_____

15. G_____

16. H_____

Objective 2 **Identify the four quadrants and determine which points lie within each one.**

Identify the quadrant in which each point is located.

17. (−3, 2)

17._____

18. (−1, −8)

18._____

19. (0, 5)

19._____

20. (−10, 2)

20._____

21. (3, −9)

21._____

22. (−3, 0)

22._____

23. (–2, –2)

23._____

24. (12, 4)

24._____

Complete each ordered pair with a number that will make the point fall in the specified quadrant.

25. Quadrant III (–2, ____)

25._____

26. No quadrant (6, ____)

26._____

27. Quadrant I (____, 3)

27._____

28. No quadrant (____, –8)

28._____

29. Quadrant II (–3, ____)

29._____

30. Quadrant IV (12, ____)

30._____

9.5   Introduction to Graphing Linear Equations

Section 9.5 Objectives
1. Graph linear equations in two variables.
2. Identify the slope of a line as positive or negative.

Key Terms *Answer the following questions about the key terms for Section 9.5.*

graph a linear equation

1. When **graphing a linear equation**   1._____
        a. plot at least three of the
        ordered pairs and connect them
        with a straight line
        b. plot all of the ordered pairs
        and connect them with a
        straight line
        c. plot the one ordered pair that
        is the true solution

Objective 1   Graph linear equations in two variables.

Graph each equation. Make your own table using the listed values of x.

1. $y = 3x$ **1.**
    Use –1, 0, and 1 as the
    values of *x*.

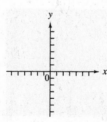

2. $y = -\dfrac{1}{4}x$ **2.**
    Use –4, 0, 4 as the
    values of *x*.

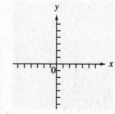

3. $y = x - 4$
Use 0, 2, and 4 as the
value of x.

3.

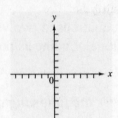

4. $y = x + 2$
Use –2, 0, and 2 as the
values of x.

4.

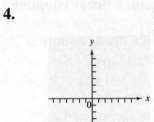

5. $x + y = -3$
Use –4, –3, and –2 as the
values of x.

5.

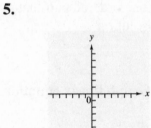

6. $x + y = 7$
Use 2, 3, and 4 as the
values of x.

6.

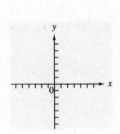

7. $y = \dfrac{1}{5}x$
Use –5, 0, and 5 as the
values of x.

7.

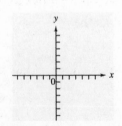

8. $x - y = 0$
Use -2, 0, and 2 as the
values of x.

8.

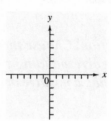

9. $y = -3x + 2$
Use 0, 1, and 2 as the
values of x.

9.

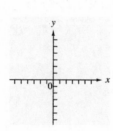

10. $y = 4x - 1$
Use -1, 0, and 1 as the
values of x.

10.

| Objective 2 | Identify the slope of a line as positive or negative.

Look back at the graphs in Problems 1–10. Then complete these sentences.

11. The graph of $y = 3x$ has a _____ slope.

12. The graph of $y = -\frac{1}{4}x$ has a _____ slope.

13. The graph of $x + y = -3$ has a _____ slope.

14. The graph of $x - y = 0$ has a _____ slope.

15. The graph of $y = -3x + 2$ has a _____ slope.

16. The graph of $y = 4x - 1$ has a _____ slope.

Objectives 1 and 2

Graph each equation. Choose three values for x. Make a table showing your x values and the corresponding y values. After graphing the equation, state whether the line has a positive or negative slope.

17. $y = -2x + 5$

17.

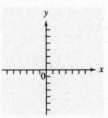

18. $y = x - 2$

18.

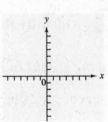

19. $x + y = 8$

19.

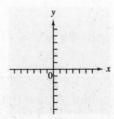

20. $y = -\dfrac{1}{2}x$

20.

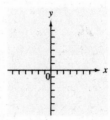

21. $y = 6x$

21.

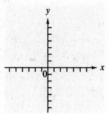

22. $x - y = 5$

22.

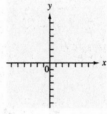

23. $y = 4x - 5$

23.

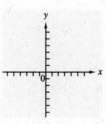

24. $y = x + 3$

24.

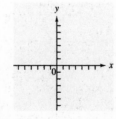

Chapter 10   EXPONENTS AND POLYNOMIALS

10.1   The Product Rule and Power Rules for Exponents

Section 10.1 Objectives

1. Review the use of exponents.
2. Read and interpret data from a pictograph.
3. Use the rule $(a^m)^n = a^{mn}$.
4. Use the rule $(ab)^m = a^m b^m$.
5. Use the rule $\left(\dfrac{a}{b} \right)^m = \dfrac{a^m}{b^m}$.

Objective 1   **Review the use of exponents.**

Write each expression using exponents.

1. **(a)** $x \times x \times x \times x \times x \times x$ 1. (a)_____

   (b) $(-8)(-8)(-8)$ (b) _____

   (c) $(ab)(ab)(ab)(ab)$ (c) _____

2. **(a)** $\dfrac{2}{5} \times \dfrac{2}{5} \times \dfrac{2}{5} \times \dfrac{2}{5} \times \dfrac{2}{5} \times \dfrac{2}{5}$ 2. (a)_____

   (b) $\left(-\dfrac{1}{9} \right)\left(-\dfrac{1}{9} \right)$ (b) _____

   (c) $(3c)(3c)(3c)$ (c) _____

Evaluate each expression.

3. **(a)** 2^6 3. (a)_____

   (b) $(-5)^3$ (b) _____

   (c) -3^4 (c) _____

4. (a) $(-1)^7$

    (b) -2^4

    (c) 11^2

4. (a)_____

(b) _____

(c) _____

Objective 2   Use the product rule for exponents.

Use the product rule to simplify each expression. Write answers in exponential form.

5. $5^3 \times 5^5$

5._____

6. $(-5)^4(-5)^5$

6._____

7. $h^2 \cdot h^{11} \cdot h$

7._____

8. $x^6 \times x^2 \times x^5$

8._____

9. $(5r^4)(-2r^3)$

9._____

10. $(-2c^7)(-4c^8)$

10._____

Objective 3   Use the rule $(a^m)^n = a^{mn}$.

Simplify each expression. Write answers in exponential form

11. $(7^3)^4$

11._____

12. $(6^3)^3$

12._____

13. $(t^6)^7$

13._____

14. $(q^5)^6$

14._____

15. $-(v^4)^9$

15._____

16. $3 \times (y^5)^4$

16._____

Objective 4   Use the rule $(ab)^m = a^m b^m$.

Simplify each expression.

17. $(4w)^4$

17._____

18. $(5u)^3$

18._____

19. $2(xz)^7$

19._____

20. $4(kn)^6$

20._____

21. $3(5mn)^2$

21._____

22. $5(3xy)^4$

22._____

Name: Date:

Instructor: Section:

Objective 5   **Use the rule** $\left(\dfrac{a}{b}\right)^m = \dfrac{a^m}{b^m}$.

Simplify each expression. Write answers in exponential form. Assume all variables represent nonzero real numbers.

23. $\left(\dfrac{7}{3}\right)^3$

23._____

24. $\left(\dfrac{2}{9}\right)^5$

24._____

25. $\left(\dfrac{6}{5}\right)^2$

25._____

26. $\left(\dfrac{c}{3y}\right)^7$

26._____

27. $\left(\dfrac{m}{g}\right)^6$

27._____

28. $\left(\dfrac{u}{p}\right)^{11}$

28._____

29. $\left(\dfrac{8}{y}\right)^4$

29._____

30. $\left(\dfrac{z}{16}\right)^3$

30._____

10.2   Integer Exponents and the Quotient Rule

Section 10.2 Objectives
1. Use 0 as an exponent.
2. Use 0 as an exponent.
3. Use the quotient rule for exponents.
4. Use the product rule with negative exponents.

Objective 1   Use 0 as an exponent.

Evaluate each expression. Assume that all variables represent nonzero real numbers.

1. (a) 3^0 1. (a)_____

    (b) -7^0 (b) _____

    (c) y^0 (c) _____

2. (a) 12^0 2. (a)_____

    (b) $(-3c)^0$ (b) _____

    (c) $-2a^0$ (c) _____

Objective 2   Use negative numbers as exponents.

Evaluate or simplify each expression, and write it using only positive exponents. Assume that all variables represent nonzero real numbers.

3. 6^{-2} 3._____

4. 9^{-5} 4._____

5. x^{-7}

_____

6. b^{-5}

6._____

7. $2^{-1} - 5^{-1}$

7._____

8. $4^{-1} + 3^{-1}$

8._____

Objective 3   **Use the quotient rule for exponents.**

Use the quotient rule to simplify each expression, and write it using only positive exponents. Assume that all variables represent nonzero real numbers.

9. $\dfrac{7^5}{7^3}$

9._____

10. $\dfrac{4^9}{4^2}$

10._____

11. $\dfrac{2^{11}}{2^{12}}$

11._____

12. $\dfrac{9^4}{9^9}$

12._____

13. $\dfrac{4^3}{4^3}$

13._____

14. $\dfrac{10^{-2}}{10}$

14._____

15. $\dfrac{5^{-5}}{5^{-3}}$

15._____

16. $\dfrac{m^{-3}}{m^{-8}}$

16._____

17. $\dfrac{t^{-7}}{t^{5}}$

17._____

18. $\dfrac{13^{4}}{13^{-3}}$

18._____

19. $\dfrac{3^{-10}}{3^{10}}$

19._____

20. $\dfrac{s}{s^{-9}}$

20._____

Objective 4   **Use the product rule with negative exponents.**

Use the product rule to simplify each expression, and write it using only positive exponents. Assume that all variables represent nonzero real numbers

21. $9^{5}(9^{-3})$

21._____

22. $3^{-9}(3^{4})$

22._____

23. $b^{-3}(b^2)$

23._____

24. $z^{10}(z^{-8})$

24._____

25. $(4^{-5})(4^{-5})$

25._____

26. $(6^{-2})(6^{-1})$

26._____

27. $(k^{-4})(k^{-4})$

27._____

28. $(u^{-9})(u^4)(u^{-2})$

28._____

29. $(m^{-1})(m^{10})(m^{-5})$

29._____

30. $p^{-7} \times p^4 \times p^{15}$

30._____

10.3   An Application of Exponents: Scientific Notation

Section 10.3 Objectives
1. Express numbers in scientific notation.
2. Convert numbers in scientific notation to numbers without exponents.
3. Use scientific notation in calculations.
4. Solve application problems using scientific notation.

Key Terms: *Answer the following questions about the key terms for Section 10.3*

scientific notation

1. Using $a \times 10^n$ is a way to

    a. express very large or very small
      numbers

    b. simplify an expression

    c. solve an equation

    d. find unknown exponents

1._____

Objective 1   **Express numbers in scientific notation.**

Write each number in scientific notation.

1. 27,500

1._____

2. 784,000

2._____

3. 38,600,000

3._____

4. 9,540,000

4._____

5. 0.0503

5._____

6. 0.0002208

6._____

7. 0.007068

7._____

8. 0.00000476

8._____

| Objective 2 | Convert numbers in scientific notation to numbers without exponents. |

Write each number without exponents.

9. 2.84×10^6

9._____

10. 7.51×10^{-5}

10._____

11. 9.331×10^{-2}

11._____

12. 5.881×10^6

12._____

13. 7.6×10^{-4}

13._____

14. 3.65×10^{-3}

14._____

15. 2.0052×10^3

15._____

16. 1.0302×10^4

16._____

Objective 3   **Use scientific notation in calculations.**

Perform the indicated operations, and write the answers in scientific notation

17. $(7 \times 10^5)(3 \times 10^{-2})$ **17.**_____

18. $(4 \times 10^{-7})(5 \times 10^4)$ **18.**_____

19. $\dfrac{8 \times 10^5}{2 \times 10^{11}}$ **19.**_____

20. $\dfrac{9 \times 10^6}{1 \times 10^{-2}}$ **20.**_____

21. $(3.5 \times 10^{-8})(2.4 \times 10^{-11})$ **21.**_____

22. $(1.3 \times 10^{-9})(8 \times 10^5)$ **22.**_____

23. $\dfrac{\left(27.2\times10^{9}\right)\left(2.5\times10^{-4}\right)}{3.2\times10^{2}}$

23._____

24. $\dfrac{3.5\times10^{4}}{\left(6.25\times10^{5}\right)\left(2.8\times10^{-8}\right)}$

24._____

25. $\dfrac{\left(6\times10^{-3}\right)\left(8\times10^{-2}\right)}{\left(2\times10^{-6}\right)\left(3\times10^{3}\right)}$

25._____

26. $\dfrac{\left(7.1\times10^{3}\right)\left(1.2\times10^{-8}\right)}{\left(3\times10^{-5}\right)\left(2\times10^{7}\right)}$

26._____

Objective 4   Solve application problems using scientific notation.

Solve each problem and write the answers in both scientific notation and without exponents.

27. There are about 6×10^{23} atoms in a mole of atoms. How many atoms are there in 8.1×10^{-15} mole?

27._____

28. There are about 93 people per square mile living in Wisconsin. The state has an area of 6.55×10^4 square miles. What is the population of Wisconsin?

28._____

29. In one minute, light travels about 1.12×10^7 miles. How far will light travel in one second?

29._____

30. Corporation A has assets of 3.9×10^{10} dollars. Corporation B has assets of 7.8×10^{12} dollars. The assets of Corporation B are how many times greater than Corporation A's assets?

30._____

10.4   Adding and Subtracting Polynomials

Section 10.4 Objectives
1. Review combining like terms.
2. Use the vocabulary for polynomials.
3. Evaluate polynomials.
4. Add polynomials.
5. Subtract polynomials.

Key Terms: *Answer the following questions about the key terms for Section 10.3*

polynomial	descending powers	degree of a term
degree of a polynomial	monomial	binomial

trinomial

1. $2x^3 - x^2 + \dfrac{4}{x}$ is NOT a **polynomial** because   1._____

    a. there is a variable in the denominator

    b. there is a coefficient in $2x^3$

    c. the variable in $\dfrac{4}{x}$ has no exponent

    d. there are too many terms

2. Descending powers means   2._____

    a. the exponents on x become smaller from right left

    b. the exponents on x may not be larger than 3

    c. the exponents on x become smaller from left to right

    d. the coefficients become smaller from left to right

3. To find the **degree of the term** $3x^4 + x^2 - x$ 3._____

    a. add $4 + 2$ to get degree 6

    b. remember that x has an understood exponent of 1 and get degree 7

    c. add $4 + 2$ and subtract the exponent of 1 to get degree 5

    d. count the number of variables and get degree 1

4. The **degree of a polynomial**   4._____

    a. cannot be found in a monomial

    b. can only be calculated if exponents are larger than 1

    c. is the number of terms in a polynomial

    d. is the highest degree of any term of the polynomial

Name: Date:
Instructor: Section:

Match the term to the correct
polynomial

a. $4y^5 + y^3 - 3$

b. $-5x^3$

_____5. monomial

_____6. binomial

c. $3m^2 - 9m$

_____7. trinomial

Objective 1   Review combining like terms.

Simplify each polynomial when possible. Write the result with descending powers.

1. $5y^9 - 11y^9$

1._____

2. $-7a^3 - 4a^{13}$

2._____

3. $-1.3z^7 + 0.4z$

3._____

4. $0.7b^4 - 8b^2$

4._____

5. $2v^3 - 7v^3 + 4v^2 + 8v$

5._____

6. $-6c^4 - 6c^2 + 9c^4 - 4c^2$

6._____

Objective 2   Use the vocabulary for polynomials.

7. In the term $-4a^7$,

7.

the coefficient is **(a)**_____ ,

(a)_____

the exponent is **(b)**_____,

(b)_____

and the degree of the term is
        (c)_____.

(c)_____

8. The degree of the polynomial

$-5x^2 + 2x^4$ is

(a)_____ and it

(b) _____

     is/is not

written in descending powers.

8.

(a)_____

(b)_____

State whether each of the following polynomials is a **monomial, binomial, trinomial,** *or* **none of these.**

9. (a) $t^3 - 6t^2 - 9$

   (b) $2y^3 - 9$

9. (a)_____

  (b)_____

10. (a) $r^4 - 2r^3 + 7r - 11$

   (b) $-122w^{11}$

10. (a)_____

  (b)_____

Objective 3   **Evaluate polynomials.**

Find the value of each polynomial (a) when x = – 2 and (b) when x = 3.

11. $-3x^2 + 2x - 5$

11.(a)_____

  (b)_____

12. $x^3 - 6$

12.(a)_____

  (b)_____

13. $3x^3 + 4x - 19$

13.(a)_____

(b)_____

14. $-4x^3 + 10x^2 - 1$

14.(a)_____

(b)_____

Objective 4   **Add polynomials.**

Add.

15. $5b^3 - 5b^2$ and $4b^3 - 7b^2$

15._____

16. $8n^2 + 2n$ and $19n^2 - 5n$

16._____

17. $-7c^2 - 2c + 3$
$\underline{\quad 4c^2 + 8c - 1}$

17._____

18. $9m^3 - m + 11$
$\underline{-2m^3 - 7m - 4}$

18._____

19. $(x^2 + 6x - 8) + (3x^2 - 10)$ **19.** _____

20. $(z^3 - 3z + 11) + (-7z^2 + 5z)$ **20.** _____

21. $(3r^3 + 5r^2 - 6) + (2r^2 - 5r + 4)$ **21.** _____

22. $(y^2 - y + 8) + (3y^3 - y - 10)$ **22.** _____

| Objective 5 | Subtract polynomials. |

Subtract.

23. $(12z^4 - 2z^2) - (4z^4 + 3z^2)$ **23.** _____

24. $(10y^5 - y^3) - (7y^5 - y^3)$ **24.** _____

25. Subtract $s^2 + 2s + 1$ from $4s^2 + 9s - 2$

25._____

26. Subtract $9t^2 + 15t - 3$ from $-2t^2 + 11t + 4$

26._____

27. $(-8w^3 + 11w^2 - 12) - (-10w^2 + 3)$

27._____

28. $(-6v^3 + 11v - 5) - (9v^3 + 11v)$

28._____

29. $(5a^4 - 6a^2 + 9a) - (a^3 - 19a - 1)$

29._____

30. $(8b^4 - 4b^3 + 7) - (2b^2 + b + 9)$

30._____

10.5   Multiplying Polynomials: An Introduction

Section 10.5 Objectives
1. Multiply a monomial and a polynomial.
2. Multiply two polynomials.

Objective 1   Multiply a monomial and a polynomial.

Find each product.

1. $8a(a-1)$

1._____

2. $12x(-9x^2+10)$

2._____

3. $9c^2(2c-5)$

3._____

4. $-9h^2(-12h^2-9h)$

4._____

5. $-11n^3(7n^2+4)$

5._____

6. $3p(-2p^2+6p+1)$

6._____

7. $-6r^2(5r^2-4r+9)$

7._____

8. $5t^3(8t^3-5t^2+9)$

8._____

9. $-4s^2(-12s^3+7s^2-10s-2)$

9._____

10. $-10q^4(3q^4+9q^2-2q+6)$

10._____

11. $(-3k^3+10)(-5k^3)$

11._____

12. $11b^2(-7b^3+6)$

Objective 2   **Multiply two polynomials.**

Find each product.

13. $(k+4)(k-12)$

13._____

14. $(m-5)(m+8)$

14._____

15. $(10b-7)(9b+11)$

15._____

16. $(6k-10)(-6k+5)$

16._____

17. $(2v+3)(6v+7)$

17._____

18. $(9w+4)(5w^2 - 4w+12)$

18._____

19. $(-7x+9)(12x^2 + x+10)$

19._____

20. $(-6d^2 +2d - 11)(2d^2 +14)$

20._____

21. $(-7u+4)(9u^3 - 10u^2 +u+2)$

21._____

22. $(4z^2 + 2z - 11)(3z^2 - 12z - 8)$

22._____

23. $(x - 2)(x + 12)$

23._____

24. $(-9d + 5)(d - 2)$

24._____

25. $(m^2 + 7m - 8)(-7m - 4)$

25._____

26. $(6w^2 - 9w - 8)(8w^2 + 10w + 1)$

26._____

Chapter R   WHOLE NUMBERS REVIEW

R.1   Adding Whole Numbers

Section R.1 Objectives
1. Add two or three single-digit numbers.
2. Add more than two numbers.
3. Add when regrouping is not required.
4. Add with regrouping.
5. Use addition to solve application problems.
6. Check the sum in addition.

Key Terms: *Answer the following questions about the key terms for Section R.1.*

addends   sum (total)   commutative property of addition

associative property of addition   carrying (regrouping)

1. When you **add** numbers together, the   1._____
   answer is called the
   a. sum
   b. quotient
   c. product
   d. opposite

2. The definition of the **commutative**   2._____
   property of addition is
   a. Changing the grouping of addends
      does not change the sum.
   b. Changing the order of two factors
      does not change the product.
   c. Changing the order of two
      addends does not change the sum.
   d. Changing the grouping of factors
      does not change the product.

3. Regrouping in **addition** is used when   3._____
   a. A digit is less than the one below
      it
   b. The sum of the digits exceeds 9

4. The example, $(5 + 8) + 10 = 5 + (8 + 10)$, illustrates which property?
   a. the **distributive property.**
   b. the **associative property of addition.**
   c. the **associative property of multiplication.**
   d. the **commutative property of addition.**

4._____

5. The number 6 is an **addend** in which problem below?
   a. $4 + 27 = 6$
   b. $6 - 3 = 3$
   c. $5 + 6 = 11$
   d. $6 \div 2 = 3$

5._____

Objective 1   Add two or three single-digit numbers.

Add.

1. $0 + 5$

1._____

2. $3 + 5$

2._____

3. $7 + (7 + 3)$

3._____

4. $(6 + 4) + 9$

4._____

Objective 2   Add more than two numbers.

Add.

5.	3	6.	8	7.	7	8.	3
	5		9		1		2
	7		3		2		9
	2		4		7		8
	+ 1		+ 2		+ 6		+ 6

Objective 3   Add when regrouping is not required.

Add

9.	2730	10.	64,052
	+ 1240		+ 23,145

Objective 4   Add with regrouping.

Add by using regrouping.

11.	86	12	64	13.	26
	+ 49		+ 87		+ 98

14.	44	15.	742	16.	715
	+ 59		+ 429		+ 592

17.   7319
           6488
         + 4677

---

18.   95
           4628
          + 328

---

19.   23
            547
           6128
          + 428

---

20.   6519
         + 4787

---

21.   358
          + 6732

---

22.   5126
              9
          + 579

Objective 5   **Use addition to solve application problems.**

Solve each application problem.

23. On the first floor of a three-floor
office building there are 410
employees. On the second floor
there are 165 employees, while on
the third floor there are 78
employees. How many employees
work in the building?

23._____

24. At the hardware store, Nate bought a
hammer that cost $15, $6 worth of
screws, and $8 worth of nails. How
much did Nate spend together?

24._____

25. Bob brought 12 golf clubs to the golf
course while Julie brought 15 golf
clubs. How many golf clubs did Bob
and Julie bring in all?

25._____

26. Sally has 587 baseball cards while Tonto has 4,956 baseball cards. What total number of cards do the two friends have?

26._____

Objective 6   Check the sum in addition.

Check each sum by adding from bottom to top. If the answer is incorrect, find the correct sum.

27. _____	28. _____	29. _____	30. _____
286	18	3429	4319
317	375	289	48
+ 455	27	34	513
1058	+ 535	+ 6227	+ 79
	956	9967	4959

R.2   Subtracting Whole Numbers

Section R.2 Objectives
1. Change addition problems to subtraction problems or the reverse.
2. Identify the minuend, subtrahend, and difference.
3. Subtract when no regrouping is needed.
4. Use addition to check subtraction answers.
5. Subtract with borrowing.
6. Use subtraction to solve application problems.

Key Terms: Answer the following questions about the key terms for Section R.1.

minuend   **subtrahend**   **difference**   **borrowing(regrouping)**

1. Regrouping in **subtraction** is used when   1._____
    a. a digit is less than the one below it
    b. the sum of the digits exceeds 9

2. The **minuend** is 5 in which problem below?   2._____
    a. $12 - 7 = 5$
    b. $5 - 1 = 4$
    c. $6 - 5 = 1$
    d. $5 + 3 = 8$

3. The **subtrahend** is 5 in which problem   3._____
below?
    a. $12 - 7 = 5$
    b. $5 - 1 = 4$
    c. $6 - 5 = 1$
    d. $5 + 3 = 8$

4. The **difference** is 5 in which problem   4._____
below?
    a. $12 - 7 = 5$
    b. $5 - 1 = 4$
    c. $6 - 5 = 1$
    d. $5 + 3 = 8$

Objective 1 Change addition problems to subtraction problems or the reverse.

Write two subtraction problems for each addition problem.

1. $2 + 6 = 8$

    1._____

    _____

2. $3 + 9 = 12$

    2._____

    _____

Write an addition problem for each subtraction problem.

3. $32 - 23 = 9$

    3._____

4. $74 - 51 = 23$

    4._____

Objective 2 Identify the minuend, subtrahend, and difference.

5. Identify the minuend, subtrahend, and difference in Exercise 3.

    5._____

    _____

    _____

6. Identify the minuend, subtrahend, and difference in Exercise 4.

    6._____

    _____

    _____

Objective 3 — Subtract when no regrouping is needed.

Subtract.

7.   485
   − 265

    _____

8.   672
   - 540

    _____

9.   4327
   - 115

    _____

10.   3789
   − 287

    _____

Objective 4 — Use addition to check subtraction answers.

Use addition to check each answer. If incorrect, find the correct answer.

11.   87
   - 24
    53

    11._____

12.   64
   - 41
    23

    12._____

13.   6541
   - 3421
    3120

    13._____

14.   397
   - 224
    183

    14._____

Objective 5   **Subtract with regrouping.**

Subtract.

15   57
       - 46
_____

16.   84
        - 39
_____

17.   730
        - 298
_____

18.   320
        - 128
_____

19.   741
        - 29
_____

20.   157
        - 98
_____

21.   8743
        - 934
_____

22.   2437
        - 1448
_____

23.   52,306
        - 4,667
_____

24.   79,000
        - 42,897
_____

25.   45,000
        -   345
_____

26.   76,015
        -    83
_____

Objective 6   Use subtraction to solve application problems.

Solve each application problem.

27. On Saturday, Ben spent $184. On Sunday he spent $46. How much more did Ben spend on Saturday?

27._____

28. Elise bowled 79 in her first game. In her second game, she bowled 146. By how many pins did Elise improve her score in the second game?

28._____

29. At a particular stadium, 2351 people attended for a high school football game on Friday. On Saturday, 15,709 attended the college football game. How many more people attended on Saturday than on Friday?

29._____

30. According to mining records, the Lost Dutchman mine yielded a total of $7,460,390 in gold. The Yellow Jacket mine yielded $4,230,560 in gold. How much more did the Lost Dutchman mine yield than the Yellow Jacket?

30._____

R.3   Multiplying Whole Numbers

Section R.3 Objectives
1. Identify the parts of a multiplication problem.
2. Do chain multiplications.
3. Multiply by single-digit numbers.
4. Multiply quickly by numbers ending in zeros.
5. Multiply by numbers having more than one digit.
6. Use multiplication to solve application problems.

Key Terms: Answer the following questions about the key terms for Section R.3.

factors   **product**   **commutative property of multiplication**

associative property of multiplication   **multiple**

1. In addition, **addends** are the numbers being added together. In multiplication, _____ are the numbers being multiplied together.
   a. Factors
   b. Sums
   c. Products
   d. Divisors

1._____

2. In $6 \times 5 = 30$, 30 is
   a. the **product** and a **multiple**
   b. the **quotient** and a **multiple**
   c. a **factor** and the **sum**
   d. a **divisor** and the **quotient**

2._____

3. The answer to a multiplication problem is called a _____.

3._____

4. Rewriting the example 5(4 · 1) as (5 · 4)1 is an example of which property?
   a. The **distributive property.**
   b. The **associative property of multiplication.**
   c. The **commutative property of multiplication**.
   d. The **associative property of addition**.

4._____

5. Rewriting the example $5 \cdot 9$ as $9 \cdot 5$ is   **5.**_____

an example of which property?

    a. The **distributive property.**

    b. The **associative property of multiplication.**

    c. The **commutative property of multiplication.**

    d. The **associative property of addition**.

Objective 1   **Identify the parts of a multiplication problem.**

1. $3 \times 5 = 15$

    1.

    factors_____

    product_____

2. $(9)(7) = 63$

    2.

    factors_____

    product_____

Objective 2   **Do chain multiplications.**

Find each product. Try to do the work mentally.

3. **(a)** $3 \times 5 \times 8$ **3. (a)**_____

    (b) $5 \cdot 7 \cdot 0$ **(b)**_____

    (c) $2 \cdot 1 \cdot 9$ **(c)**_____

    (d) $(4)(3)(8)$ **(d)**_____

4. **(a)** $4 \times 6 \times 7$ **4. (a)**_____

    (b) $8 \cdot 0 \cdot 4$ **(b)**_____

    (c) $7 \cdot 3 \cdot 8$ **(c)**_____

    (d) $(6)(2)(1)$ **(d)**_____

Objective 3   **Multiply by single-digit numbers.**

Multiply.

5.   612
   ×   8
   _____

6.   733
   ×   5
   _____

7.   4178
   ×   3
   _____

8.   5276
   ×   4
   _____

9.   17,251
   ×   9
   _____

10.   24,542
    ×   7
   _____

Objective 4   **Multiply quickly by numbers ending in zeros.**

Multiply.

11.   327
    × 40
   _____

12.   637
    × 50
   _____

13.   500
    × 300
   _____

14.   700
    × 800
   _____

15.   42,000
    × 5,000
   _____

16   18,000
    × 7,000
   _____

Objective 5   **Multiply by numbers having more than one digit.**

17. (78)(25)

18. (63)(38)

18._____

19. (315)(64)

19._____

20. (236)(58)

20._____

21.   473
        × 189
    _____

22.   623
        × 421
    _____

23.   2358
        × 705
    _____

24.   4671
        × 309
    _____

Objective 6   Use multiplication to solve application problems.

Solve each application problem. You may need to use addition and subtraction as well as multiplication.

25. Carol works an average of 8 hours per day at her job. If in a given year she worked a total of 236 days, how many total hours did Carol work?

25._____

26. A grocery store stocked 42 bottles of one particular brand of vitamin C. There were 325 tablets in each bottle. How many tablets of this brand of vitamin C does the store have in all?

26._____

27. Bernadette types 76 words per minute while Jacki types 83 words per minute. In 20 minutes of typing, how many more words does Jacki type than Bernadette?

27._____

28. Velma's car gets 32 miles per gallon on the highway. How many miles can Velma drive on 13 gallons of gas?

28._____

29. In one apple orchard, an average tree produces 200 apples per week at the peak of the growing season. How many apples would be produced during such a week if the orchard has 174 apple trees?

29._____

30. Vincent bought 5 shirts for $22 each, 4 pairs of pants for $34 each, and 8 pairs of socks for $6 each. Find the total amount he spent.

30._____

R.4   Dividing Whole Numbers

Section R.4 Objectives
1. Write division problems in three ways.
2. Identify the parts of a division problem.
3. Divide 0 by a number.
4. Recognize that division by 0 is undefined.
5. Divide a number by itself.
6. Use short division.
7. Use multiplication to check quotients.
8. Use tests for divisibility.

Key Terms: *Answer the following questions about the key terms for Section R.4.*

dividend   divisor   quotient   short division   remainder

Match the correct word to its number in the following division problem:

$$\begin{array}{r} 450 \quad R65 \\ 72\overline{)32,465} \end{array}$$

_____**1**. Quotient   **a**. 65

_____**2**. Divisor   **b**. 450 R

_____**3**. Dividend   **c**. 32,465

_____**4**. Remainder   **d**. 72

5. In the problem $48 \div 6 = 8$, you would use   **5**._____
      a. long division
      b. short division

Objective 1   **Write division problems in three ways.**

Rewrite each division using two other symbols.

    1. (a) $52 \div 4 = 13$ **1. (a)**_____

    (b) $19\overline{)38}^{\,2}$ **(b)**_____

2. (a) $27 \div 9 = 3$

2. (a)_____

(b) $\dfrac{56}{7} = 8$

(b)_____

Objective 2   **Identify the parts of a division problem.**

Identify the dividend, divisor, and quotient.

3. (a) $8 \div 4 = 2$

3.(a)_____

_____

_____

3. (b) $\dfrac{48}{16} = 3$

3.(b)_____

_____

_____

4. (a) $36 \div 4 = 9$

4.(a)_____

_____

_____

4. (b) $8\overline{)72}$ with quotient 9

4.(b)_____

_____

_____

Objective 3 Divide 0 by a number.

Objective 4 Recognize that division by 0 is undefined.

Objective 5 Divide a number by itself.

Divide.

5. $0 \div 4$ _____

6. $33\overline{)33}$ _____

7. $\dfrac{3}{0}$ _____

8. $\dfrac{0}{17}$ _____

9. $\dfrac{16}{16}$ _____

10. $39 \div 0$ _____

11. $48\overline{)0}$ _____

12. $0\overline{)678}$ _____

Objective 6 Use short division.

Divide by using short division.

13. $5\overline{)210}$ _____

14. $6\overline{)138}$ _____

15. $9945 \div 7$ _____

16. $\dfrac{5607}{8}$ _____

17. $2803 \div 5$ _____

18. $4\overline{)2080}$ _____

19. $1014 \div 4$ _____

20. $\dfrac{549}{9}$ _____

21. $\dfrac{26,207}{3}$ _____

22. $13,573 \,\overline{\smash{)}\,} 5$ _____

23. $\dfrac{323,742}{6}$ _____

24. $28,370 \div 7$ _____

Objective 7   **Use multiplication to check quotients.**

Check each quotient. If a quotient is incorrect, find the correct quotient.

25. $4\overline{\smash{)}\,231,406}$ quotient $57,850$ R2

25. _____

26. $3\overline{\smash{)}\,23,301}$ quotient $7\ 767$

26. _____

27. $6\overline{\smash{)}\,27,014}$ quotient $4\ 502$ R1

27. _____

28. $7\overline{\smash{)}\,2304}$ quotient 329 R

28. _____

Objective 8   Use tests for divisibility.

Use the divisibility test to decide whether the number is divisible by 2, 3, 5, or 10.

29. (a)   700   29.(a)_____

    (b)   153   (b)_____

30. (a)   60   30.(a)_____

    (b)   75   (b)_____

R.5   Long Division

Section R.5 Objectives
1. Do long division.
2. Divide multiples of 10.
3. Use multiplication to check quotients.

Key Terms: *Answer the following questions about the key terms for Section R.5.*

long division

1. The problem $42\overline{)23,730}$ (with 565 above) requires long

division because

      a. The dividend has more than one digit

      b. The quotient has more than one digit

      c. The divisor has more than one digit

      d. There is no remainder

1._____

Objective 1   Do long division.

Divide by using long division.

1. $31\overline{)6659}$

1._____

2. $24\overline{)1344}$

2._____

3. $35\overline{)2193}$

3._____

4. $53\overline{)27,654}$

4. _____

5. $37\overline{)11,026}$

5. _____

6. $85\overline{)47,697}$

6. _____

7. $61\overline{)342,827}$

7. _____

8. $52\overline{)328,692}$

8. _____

9. $44\overline{)229,437}$

9. _____

10. $235\overline{)296,335}$

10._____

11. $629\overline{)159,693}$

11._____

12. $437\overline{)421,445}$

12._____

Objective 2   **Divide multiples of 10.**

Divide.

13. $70 \div 10$

13._____

14. $1600 \div 100$

14._____

15. $180,000 \div 100$

15._____

16. $275,000 \div 1000$

16._____

17. $30\overline{)1200}$

17._____

18. $50\overline{)3500}$

_____

19. $220\overline{)84,700}$

19._____

20. $170\overline{)111,010}$

20._____

21. $5400\overline{)32,400}$

21._____

22. $2100\overline{)113,400}$

22._____

Objective 3   **Use multiplication to check quotients.**

Check each division. If a quotient is incorrect, find the correct quotient.

23. $33\overline{)7210}$ $\overset{218\ \text{R}14}{}$

23._____

24. $29\overline{)5431}$ $\overset{185\ \text{R}8}{}$

24._____

25. $47\overline{)6284}$ $\overset{133\ \text{R}33}{}$

25._____

26. $51\overline{)56,214}$ $\overset{1,102\ \text{R}8}{}$

26._____

$$\begin{array}{r} 1{,}753 \ \text{R5} \\ 15\overline{)26{,}315} \end{array}$$

27.

27._____

$$\begin{array}{r} 1{,}034 \ \text{R9} \\ 19\overline{)19{,}653} \end{array}$$

28.

28._____

$$\begin{array}{r} 98 \ \text{R534} \\ 653\overline{)64{,}528} \end{array}$$

29.

29._____

$$\begin{array}{r} 143 \ \text{R161} \\ 341\overline{)48{,}926} \end{array}$$

30.

30._____

INTRODUCTION TO ALGEBRA: INTEGERS

1.1 Key Terms

1. A
3. B

1.1   Place Value

Objective 1

1. 4; 0

Objective 2

3. tens

5. ten-thousands

7. hundred-millions

9. ten millions

11. hundred billions, millions, ten thousands, hundreds

Objective 3

13. four thousand, six hundred forty

15. one hundred sixty thousand, one hundred eighty

17. twenty million, five hundred eight thousand, four hundred seventy

19. nine billion, six hundred seventy-one million, six hundred thirty-seven

21. four hundred sixty-four billion, one hundred ten million, fifty-four thousand

23. 35,096

25. 7,946,002

27. 900,011,500

29. 109,966,000,000

1.2 Key Terms

1. C
3. Absolute Value

1.2   Introduction to Signed Numbers

Objective 1

1. $^-14$ meters

3. $^-\$1.2$ million

5. $^-2$ strokes

7. $^-3$ decibels

Objective 2

9.

Objective 3

11. >

13. <

15. >

17. <

19. <

21. >

23. >

Objective 4

25. 3

27. 95

29.   926

1.3 Key Terms

1. A
3. A

1.3　Adding Integers

Objective 1

1. 2

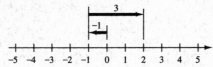

3. ⁻14

5. ⁻74

7. ⁻2

9. ⁻9

11. ⁻32

13. ⁻28

15. 9

17. ⁻9

19. 3

21. ⁻38

23. ⁻$371 + $500 = $129

Objective 2

25. ⁻9 + ⁻2 = ⁻11
    ⁻2 + ⁻9 = ⁻11

27. (⁻8 + 8) + ⁻15 ;
    0 + ⁻15 = ⁻15

29. 67

1.4 Key Terms

1. B

1.4   Subtracting Integers

Objective 1

1. $^-8; 8 + ^-8 = 0$ 3. $1; ^-1 + 1 = 0$

Objective 2

5. 1	**9.** 20	**13.** $^-19$	**17.** 8	**21.** $^-15$
7. $^-14$	**11.** 12	**15.** 11	**19.** 4	**23.** 13

Objective 3

25. $^-18$ **29.** $^-30$

27. 0

1.5 Key Terms

1. C
3. B

1.5　Problem Solving: Rounding and Estimating

Objective 1

1. $^{-}1\underline{3}5$

3. $\underline{5},132,699$

Objective 2

5. 130

7. 7000

9. $^{-}38,000$

11. 910,000

13. 10,000,000

Objective 3

15. $^{-}\$2000$

17. $^{-}10 + {}^{-}80 = {}^{-}90;\ {}^{-}88$

19. $^{-}100 + 300 = 200;\ 199$

21. $8000 + {}^{-}5000 = 3000;\ 2957$

23. $80 + {}^{+}80 = 160;\ 158$

25. $9000 + {}^{-}1000 = 8000;\ 7162$

27. $^{-}800 + {}^{-}8000 = {}^{-}8800;\ {}^{-}9067$

29. $^{-}100 + 40 + {}^{-}20 = {}^{-}80$ meters; $^{-}91$ meters

1.6 Key Terms

1. A
3. B
5. product
7. C

1.6   Multiplying Integers

Objective 1

1. $^-3 \cdot 6$; $^-3\,(6)$ or $(^-3\,)(6)$

Objective 2

3. $^-35$

5. 39

7. 58

9. 30

11. $^-27\ ^-62$

13. 22

15. 52

17. $^-30$

19. $^-162$

21. 6

23. $^-4$

25. 0

Objective

27. (a) $8 \cdot 4 + 8 \cdot\ ^-7$; Both results are $^-24$.

27. (b) $3 \cdot (^-2 \times\ ^-9)$; Both results are 54.

Objective 4

29. Estimate: $\$8 \cdot 20 = \$160$
    Exact: $\$8 \cdot 23 = \$184$

1.7   Dividing Integers

Objective 1

1. $^-8$

3. 3

5. $^-7$

7. $^-1$

9. 8

11. undefined

13. 9

Objective 2

15. 1; a number divided by itself equals 1.

17. undefined; division by 0 is undefined.

Objective

19. 20

21. 72

23. $^-21$

Objective 4

25. Estimate: $400,000 \div 20,000 = \$20$
Exact: $364,000 \div 19,200 = \$19$

27. Estimate: $3000 \div 6 = \$500$
Exact: $3102 \div \$6 = \$517$

Objective 5

29. $1690 \div 20 = 84$ R10. 84 cases can be filled. The remainder of 10 means that 10 bottles (half a case) are left over.

1.8   Exponents and Order of Operations

Objective 1

1. (a) 8^4 ; eight to the fourth power

1. (b) 9^2 ; nine squared

Objective 2

3. 81

9. 144

5. $^-128$

11. $^-121$

7. 10,000

Objective 3

13. $^-7$

21. 17

15. $^-10$

23. 9

17. 14

25. 47

27. 96

19. $^-28$

Objective 4

29. $\dfrac{42}{^-2} = {^-21}$

31. $\dfrac{^-48}{^-3} = 16$

UNDERSTANDING VARIABLES AND SOLVING EQUATIONS

2.1 Introduction to Variables

Key Terms

    1. A
    2. D
    3. D
    5. C

Objective 1

1. z is a variable:
    4 is a coefficient.

3. h is a variable:
    $^-7$ is a constant.

5. m is a variable;
    3 is a coefficient;
    11 is a constant.

7. q is a variable; one of the 2's is a coefficient; the other 2 is a constant.

Objective 2

9. (a) $120
    (b) $170

11. (a) $10,040
    (b) $12,620

13. (a) $8250
    (b) $9734

15. (a) 13
    (b) $^-47$
    (c) 3
    (d) $^-52$

17. (a) 0
    (b) $^-8$
    (c) 24
    (d) 16

Objective 3

19. Zero added to any number equals that number.

21. Multiplication is distributive over addition.

Objective 4

23. x^3

27. $^-5c^3$

25. a^2b^4

29. $^-6z^3$

2.2   Simplifying Expressions

Objective 1

1. $9b$, b; the coefficients are 9 and 1.

Objective 2

3. $^-3v^3$

5. $2t^3$

7. $^-19c^3z^4$

9. $10p - 11ab$

11. $^-2ht + 11$

13. cannot be simplified

15. $3b^4m + 9$

17. ^-8a

19. $40ab^2$

21. ^-11zct

Objective 3

23. $2x + 10$

25. $6z - 42$

27. $^-5t - 20$

29. $^-8p + 88$

2.3 Solving Equations Using Addition

Objective

1. 12   **3.** 3

Objective 2

5. $q = 5$;   Check: $q + 6 = 11$

$$5 + 6 \overset{?}{=} 11$$

$$11 = 11$$

7. $s = 13$;   Check: $4 = s - 9$

$$4 \overset{?}{=} 13 - 9$$

$$4 = 4$$

9. $k = {}^-9$

11. $g = 0$

13. $u = {}^-4$

15. ${}^-1 + v = {}^-2$

$${}^-1 + {}^-1 \overset{?}{=} {}^-2$$

$${}^-2 = {}^-2$$

Balances.

${}^-1$ is the correct solution.

17. $n - 12 = {}^-6$

$$8 - 12 \overset{?}{=} {}^-6$$

$${}^-4 \neq {}^-6$$

Does not balance.
The correct solution is 6.

Objective 3

19. $b = 10$

21. $z = {}^-1$

23. $p = {}^-27$

25. $s = {}^-8$

27. $a = 3$

29. $t = {}^-9$

2.4   Solving Equations Using Division

Objective 1

1. $s = 5$;   Check: $5s = 25$

    $5 \times 5 \overset{?}{=} 25$

    $25 = 25$

3. $a = {}^-9$; Check: ${}^-72 = 8a$

    ${}^-72 \overset{?}{=} 8 \times {}^-9$

    ${}^-72 = {}^-72$

5. $g = {}^-5$

7. $x = {}^-4$

9. $v = 5$

Objective 2

11. $d = {}^-5$

13. $k = 19$

15. $t = 0$

17. $n = {}^-7$

19. $f = {}^-20$

21. $g = {}^-3$

23. $k = {}^-3$

Objective 3

25. $f = 10$

27. $g = {}^-5$

29. $z = {}^-15$

2.5   Solving Equations with Several Steps

Objective 1

1. $g = 3$;   Check: $^-3g - 18 = ^-27$

$^-3 \times 3 - 18 \overset{?}{=} ^-27$

$^-9 - 18 \overset{?}{=} ^-27$

$^-27 = ^-27$

3. $w = 9$;   Check: $91 = 8w + 19$

$91 \overset{?}{=} 8 \times 9 + 19$

$91 \overset{?}{=} 72 + 19$

$91 = 91$

5. $z = ^-9$

7. $z = 9$

9. $m = 0$

11. $r = 4$;   Check: $2r - 15 = 9r - 43$

$2 \times 4 - 15 \overset{?}{=} 9 \times 4 - 43$

$8 - 15 \overset{?}{=} 36 - 43$

$^-7 = ^-7$

13. $k = 1$;   Check: $^-15 + 7k = ^-9k + 1$

$^-15 + 7 \times 1 \overset{?}{=} ^-9 \times 1 + 1$

$^-15 + 7 \overset{?}{=} ^-9 + 1$

$^-8 = ^-8$

15. $g = 3$

17. $n = 2$

19. $d = ^-2$

Objective 2

21. $h = ^-4$;   Check: $5(h + 3) = ^-5$

$5(^-4 + 3) \overset{?}{=} ^-5$

$5(^-1) \overset{?}{=} ^-5$

$^-5 = ^-5$

23. $r = ^-3$;   Check: $90 = ^-10(r - 6)$

$90 \overset{?}{=} ^-10(^-3 - 6)$

$90 \overset{?}{=} ^-10(^-9)$

$90 = 90$

25. $h = ^-5$

29. $p = ^-9$

27. $h = 7$

SOLVING APPLICATION PROBLEMS

3.1   Problem Solving: Perimeter

Key Terms

1. D
3. B
5. D

Objective 1

1. $P = 28$ ft

3. $P = 40$ in.

5. $s = 26$ ft **7.** $s = 19$ in.

Objective 2

9. $P = 16$ m

11. $P = 34$ yd

13. $P = 56$ in.

15. $l = 12$ in.

17. $w = 11$ m

19. $w = 17$ ft

Objective 3

21. $P = 18$ cm

23. $P = 26$ yd

25. $P = 32$ cm

27. $P = 255$ cm

29. $? = 33$ yd

3.2   Problem Solving: Area

Key Terms

1. A

Objective 1

1. $A = 10 \text{ cm}^2$

3. $A = 21 \text{ ft}^2$

5. $w = 22$ in.

7. $l = 60$ m

Objective 2

9. $A = 4 \text{ ft}^2$

11. $A = 1089 \text{ cm}^2$

13. $s = 20$ cm

15. $s = 7$ in.

Objective 3

17. $A = 527 \text{ yd}^2$

19. $A = 460 \text{ in.}^2$

21. $h = 12$ in.

23. $h = 9$ yd

Objective 4

25. $P = 46$ in.; $A = 105 \text{ in.}^2$

27. $P = 1688$ m; $A = 169{,}984 \text{ m}^2$

29. 11 yd

3.3 Solving Application Problems with One Unknown Quantity

Objective 1

1. $x + 17$ **3.** $35 - x$ **5.** $x - 15$

7. $-2 - x$ **9.** $-47x$ **11.** $8x - 11$

13. $4x + x$

Objective 2

15. 22   **17.** -6 **19.** 6

Objective 3

21. 132 lb   **23.** 7 sticks

25. $8   **27.** 22 slices

29. 12 bananas

3.4 Solving Application Problems with Two Unknown Quantities

Objective 1

1. Debbie is 35; Lori is 30.

3. Lucas earned $35,160; his wife earned $40,160.

5. Linda paid $168.75 for her dog; she paid $56.25 for her cat.

7. One piece is 53 cm long; the other is 36 cm long.

9. One piece is 60 m long; the other is 38 m long.

11. 96 males; 41 females

13. 2 ft, 2 ft, and 4 ft

15. 18 mi

17. The length is 36 yd; the width is 12 yd.

19. The length is 31 in.; the width is 12 in.

21. $A = 36$ m^2; $P = 40$ m

RATIONAL NUMBERS: POSITIVE AND NEGATIVE FRACTIONS

4.1 Introduction to Signed Fractions
Objective 1

1. (a) $\dfrac{1}{3}$; $\dfrac{2}{3}$

1. (b) $\dfrac{3}{8}$; $\dfrac{5}{8}$

1. (c) $\dfrac{5}{2}$; $\dfrac{1}{2}$

Objective 2

3. (a) N: 3; D: 8

3. (b) N: 12; D: 5

5. $\dfrac{1}{4}, \dfrac{7}{8}, \dfrac{5}{16}; \dfrac{8}{3}, \dfrac{5}{5}, \dfrac{11}{2}$

Objective 3

7.

9.

Objective 4

11. $\dfrac{3}{5}$ 12.   0   13. $\dfrac{15}{7}$ 15. $\dfrac{3}{3}$

Objective 5

17. $\dfrac{24}{48}$ 19. $\dfrac{32}{48}$ 21. $\dfrac{8}{48}$ 23. $\dfrac{18}{48}$ 25. $-\dfrac{1}{5}$

27. Already in lowest terms   29. $-\dfrac{4}{5}$

4.2   Writing Fractions in Lowest Terms

Objective 1

1. (a)   No; 2 is a common factor.

1. (b)   Yes

1. (c)   No; 5 is a common factor

Objective 2

3. (a) $\dfrac{1}{2}$

3. (b) $-\dfrac{2}{3}$

3. (c) $\dfrac{3}{8}$

Objective 3

5. $3 \cdot 3 \cdot 3$ **7.** $3 \cdot 3 \cdot 5$

9. $3 \cdot 5 \cdot 7$ **11.** $2 \cdot 2 \cdot 2 \cdot 13$

Objective 4

13. $\dfrac{\cancel{2} \times 2 \times \overset{1}{\cancel{3}}}{\cancel{2} \times \cancel{3} \times 7} = \dfrac{2}{7}$ **17.** $\dfrac{31 \times \overset{1}{\cancel{3}} \times \overset{1}{\cancel{71}}}{31 \times 5 \times \cancel{71}} = \dfrac{3}{5}$

15. $\dfrac{\cancel{2} \times 2 \times \overset{1}{\cancel{3}} \times 3}{\cancel{2} \times \cancel{3} \times 5 \times 7} = \dfrac{6}{35}$ **19.** $\dfrac{\cancel{2} \times \overset{1}{\cancel{2}} \times 3 \times 3}{\cancel{2} \times \cancel{2} \times 2 \times 2 \times 5} = \dfrac{9}{20}$

Objective 5

21. $\dfrac{3}{7}$ **25.** $\dfrac{5}{9y}$

23. $\dfrac{5r}{6s}$ **27.** Already in lowest terms

29. Already in lowest terms

4.3 Multiplying and Dividing Signed Fractions

Objective 1

1. $-\dfrac{5}{16}$ 3. $\dfrac{9}{7}$ 5. 100   7. –4

Objective 2

9. $\dfrac{\cancel{3}\times c \times c}{\cancel{3}\times 3 \times 5} = \dfrac{c^2}{15}$

13. $\dfrac{\overset{1}{\cancel{2}}\times \overset{1}{\cancel{3}}\times \overset{1}{\cancel{x}}}{\underset{1}{\cancel{2}}\times 2 \times \underset{1}{\cancel{3}}\times 5 \times \underset{1}{\cancel{x}}\times x} = \dfrac{1}{10x}$

11. $\dfrac{\overset{1}{\cancel{2}}\times 3 \times \overset{1}{\cancel{x}}}{\underset{1}{\cancel{2}}\times 2 \times 5 \times \underset{1}{\cancel{x}}} = \dfrac{3}{10}$

Objective 3

15. $\dfrac{1}{21}$

17. $-\dfrac{2}{9}$

19. –20

Objective 4

21. $\dfrac{7b^2}{10}$

23. $\dfrac{a}{b}$

25. $6x$

Objective 5

27. 12 times

28. $4000

29. 225 women; 135 men

30. 25 hats

4.4   Adding and Subtracting Signed Fractions

Objective 1

1. 1

3. $-\dfrac{5}{7}$

5. $\dfrac{4}{y^2}$

Objective 2

7. 14

9. 12

11. 105

Objective 3

13. $\dfrac{5}{6}$

17. $\dfrac{27}{40}$

21. $\dfrac{11}{48}$

15. $\dfrac{11}{24}$

19. $\dfrac{13}{60}$

Objective 4

23. $\dfrac{5}{ab}$

25. $\dfrac{15-3n}{5n}$

27. $\dfrac{-2-yz}{z^2}$

29. $-\dfrac{ab}{6c}$

4.5 Problem Solving: Mixed Numbers and Estimating

Objective 1

1.

$-2\frac{2}{3}$ $2\frac{2}{3}$

3.

$-\frac{5}{2}$ $\frac{5}{2}$

Objective 2

5. $\dfrac{13}{5}$ **7.** $-\dfrac{51}{8}$ **9.** $4\dfrac{5}{7}$ **11.** $^{-}5\dfrac{1}{4}$

Objective 3

13. $4 \cdot 1 = 4;\ 4\dfrac{4}{21}$ **15.** $4 \div 1 = 4;\ 2\dfrac{8}{11}$ **17.** $3 \cdot 5 = 15;\ 16\dfrac{2}{3}$

Objective 4

19. $4 - 3 = 1;\ 1\dfrac{3}{4}$ **21.** $3 + 4 = 7;\ 6\dfrac{7}{12}$ **23.** $5 - 2 = 3;\ 2\dfrac{3}{4}$

Objective 5

25. $1 + 2 = 3$ inches; $3\frac{8}{15}$ inches   **29.** $3 \cdot 6 = 18$ yd; $18\dfrac{3}{4}$ yd

27. $7 - 5 = 2$ hours; $1\dfrac{5}{9}$ hours

4.6 Exponents, Order of Operations, and Complex Fractions

Objective 1

1. $\dfrac{4}{25}$

3. $\dfrac{16}{81}$

5. $\dfrac{49}{64}$

7. $-\dfrac{28}{25}$ or $-1\dfrac{3}{25}$

9. $\dfrac{1}{18}$

11. $-\dfrac{2}{3}$

Objective 2

13. $\dfrac{1}{3}$

15. 0

17. -6

19. $\dfrac{7}{9}$

21. $\dfrac{59}{6}$ or $9\dfrac{5}{6}$

Objective 3

23. 2

25. -50

27. $\dfrac{4}{9}$

29. $\dfrac{25}{36}$

4.7 Problem Solving: Equations Containing Fractions

Objective 1

1. $m = 54$

3. $x = -\dfrac{5}{16}$

5. $b = -36$

7. $m = \dfrac{2}{3}$

9. $h = \dfrac{4}{3}$ or $1\dfrac{1}{3}$

11. $n = 81$

Objective 2

13. $n = 10$

15. $r = -4$

17. $x = 16$

19. $y = 16$

21. $n = -27$

23. $c = \dfrac{11}{48}$

Objective 3

25. 20 years old

27. 32 years old

29. 74 in. tall

4.8   Geometry Applications: Area and Volume

Objective 1

1. $P = 66$ m; $A = 200$ m^2

3. $P = 27\frac{3}{10}$ in.; $A = 32$ in.^{2}

5. $P = 33\frac{1}{2}$ yd; $A = 48\frac{1}{8}$ yd^2

7. $P = 107$ ft; $A = 250$ ft^2

9. $P = 61$ cm; $A = 135$ cm^2

Objective 2

11. 96 m^3

13. $67\frac{1}{2}$ ft^3

15. $37\frac{1}{27}$ in.^{3}

17. $89\frac{1}{4}$ m^3

Objective 3

19. 255 m^3

21. 196 ft^3

RATIONAL NUMBERS: POSITIVE AND NEGATIVE DECIMALS

5.1 Reading and Writing Decimal Numbers

Key Terms

1. a

2. c

3. d

Objective 1

1. (a) $\dfrac{1}{10}$; 0.1; one tenth

   (b) $\dfrac{63}{100}$; 0.63; sixty-three-hundredths

Objective 2

3. 3; 0; 6

5. 8; 3; 2

7. 3; 7; 5

9. 789.42

11. 0.3492

Objective 3

13. seven tenths

15. one hundred two thousandths

17. seventeen and one hundredth

19. 0.311

21. 300.039

23. 900.05

Objective 4

25. $15\dfrac{2}{5}$

27. $\dfrac{22}{25}$

29. $7\dfrac{1}{500}$

5.2   Rounding Decimal Numbers

Objectives 1 and 2

1. 24.8

3. 0.857

5. 0.49

7. 1.0

9. 673.0

11. 37

13. 74.0003

15. 79.82

17. 27

Objective 3

19. $400

21. $1.00

23. $1.04

25. $10,000

27. $276.00

29. $0.97

5.3   Adding and Subtracting Signed Decimal Numbers

Objective 1

1. 12.38

3. 79.9

5. 0.152

7. 10.564

9. 224.79

Objective 2

11. 75.57

13. −7.94

15. 66.953

17. −75.43

19. 2.9

21. 72.2049

Objective 3

23. *Estimate*: $4 + 4 + 90 = 98$
Exact: 93.228

25. *Estimate*: $10 − 30 = −20$
Exact: −14.735

27. *Estimate*:
$−300 + 400 = 100$
Exact: 115.86

29. *Estimate*:
$40 − 20 = 20$ hours
Exact: 19.15 hours

5.4 Multiplying Signed Decimal Numbers

Objective 1

1. −2.45

3. 25.1748

5. 223.44

7. 0.000078

9. $611,168.03

11. 227.164

13. −0.497211

Objective 2

15. *Estimate*: 10(5) = 50
    Exact: 52.2617

17. *Estimate*: 9(3) = 27

19. *Estimate*: 40 ×20 = 800
    Exact: 842.7796

21. *Estimate*: 20(5) = 100
    Exact: 70.8284

23. *Estimate*: 7 ×60 = 420
    Exact: 431.88

25. $592.62

27. $51.62

29. $18,760

5.5   Dividing Signed Decimal Numbers

Objective 1

  1. −17.84

  3. 21.625

  5. 8.39

  7. 0.1331

Objective 2

  9. 170.09

  11. −61.33

  13. 30.8

  15. 18,397.71

  17. $17.59

  19. $7.95

Objective 3

  21. $90 \div 2 = 45$; unreasonable; $89.17 \div 1.85 = 48.2$

  23. $3 \div 1 = 3$; reasonable

Objective 4

  25. 50.79

  27. −43.01

  29. −412

5.6   Fractions and Decimals

Objective 1

1. 1.375

3. 1.6

5. 0.05

7. 0.625

9. 10.571

11. 24.75

13. 0.818

15. 0.15

17. 4.222

Objective 2

19. 0.8, 0.8039, 0.804

21. 0.8, 0.8005, 0.85

23. 6.0069, 6.69, 6.7, 6.704

25. $\dfrac{1}{3}$, 0.35, 0.5, $\dfrac{4}{7}$

27. too little; 0.445 mg

29. less; 0.3 in.

5.7   Problem Solving with Statistics: Mean, Median, Mode, and Variability

Objective 1

   1. $37.20

   3. $43,175

   5. 850 people

   7. $389.80

Objective 2

   9. 24.3

   11. 6.0

Objective 3

   13. 27.5 liters

Objective 4

   15. 25 years

   17. 3 and 9 calls

   19. 40.1 hr; 40 hr; 40 hr and 42 hr

   21. $631.70; $692.50; no mode

   23. 28.1 students; 28.5 students; 30 students

   25. 3.15

Objective 5

   27. 38; 53; Student B

   29. 18°F; 14°F; Week J

5.8 Geometry Applications: Pythagorean Theorem and Square Roots

Objective 1

1. 100

3. 7

5. 3.606

7. 9.487

9. 8.307

11. 12.247

13. 54.772

Objective 2

15. 20 cm

17. 15 in.

19. 14.3 cm

21. 3.8 in.

23. 13 cm

25. 20.3 km

Objective 3

27. 6.7 ft

29. 3059.4 ft

5.9 Problem Solving: Equations Containing Decimals

Objective 1

  1. $h = -6.4$ **5.** $n = -4.6$

  3. $c = -11.8$ **7.** $b = 0.003$

Objective 2

  9. $y = 0.38$ **11.** $w = -8$ **13.** $m = 0$

Objective 3

  15. $m = -3.9$ **19.** $y = -14$ **23.** $y = -0.21$

  17. $h = 2.5$ **21.** $x = 29.4$ **25.** $r = -5.35$

Objective 4

  27. 50 milliliters

  29. 17 minutes

5.10 Geometry Applications: Circles, Cylinders, and Surface Area

Objective 1

1. r = 30 ft

3. d = 66 yd

Objective 2

5. C = 502.4 mm

7. C ≈ 69.1 in.

Objective 3

9. A ≈ 226.9 cm^2

11. A ≈ 18.1 in.^{2}

Objective 4

13. $C ≈ 59.7$ cm
$A ≈ 283.4$ cm^2

17. $C ≈ 28.3$ m.
$A ≈ 63.6$ m^2

15. $C ≈ 20.4$ ft
$A ≈ 33.2$ ft^2

19. 9891 ft^3

Objective 5

21. SA = 558 in.^{2}

Objective 6

23. SA ≈ 1507.2 in.^{2}

25. $V ≈ 32.1$ in.^{3}
$SA ≈ 85.7$ in.^{2}

27. $V = 7956$ mm^3
$SA = 2670$ mm^2

29. $V = 1331$ mm^3
$SA = 726$ mm^2

RATIO, PROPORTION, AND LINE/ANGLE/TRIANGLE RELATIONSHIPS

6.1   Ratios

Key Terms

1. A

Objective 1

1. $\dfrac{3}{11}$

3. $\dfrac{3}{1}$

5. $\dfrac{1}{4}$

7. $\dfrac{9}{7}$

9. $\dfrac{3}{4}$

Objective 2

11. $\dfrac{1}{2}$

13. $\dfrac{3}{2}$

15. $\dfrac{6}{1}$

Objective 3

17. $\dfrac{13}{11}$

19. $\dfrac{2}{9}$

21. $\dfrac{1}{5}$

23. $\dfrac{5}{12}$

25. $\dfrac{8}{3}$

27. $\dfrac{5}{18}$

29. $\dfrac{7}{4}$

6.2 Rates

Key Terms

1. B

3. D

Objective 1

1. $\dfrac{3 \text{ servings}}{1 \text{ person}}$

3. $\dfrac{\$9}{10 \text{ cards}}$

5. $\dfrac{7 \text{ feet}}{18 \text{ seconds}}$

7. $\dfrac{21 \text{ miles}}{1 \text{ gallon}}$

9. $\dfrac{5 \text{ packages}}{3 \text{ minutes}}$

11. $\dfrac{138 \text{ bushels}}{5 \text{ acres}}$

Objective 2

13. $147.20/day

15. $23/hour

17. $9/visit

19. 55 miles/hour

21. 0.5 pound/person

23. 50 miles/hour

Objective 3

25. 8 ounces for $1.29

27. 16 ounces for $3.39

29. Brand B

6.3   Proportions

Key Terms

 1. A

Objective 1

 1. (a) $\dfrac{\$12}{18 \text{ boxes}} = \dfrac{\$18}{27 \text{ boxes}}$

 1. (b) $\dfrac{15}{40} = \dfrac{3}{8}$

Objective 2

 3. $\dfrac{2}{9} = \dfrac{2}{9}$; true

 5. $\dfrac{5}{8} \neq \dfrac{5}{6}$; false

 7. $75 = 75$; true

 9. $630 \neq 588$; false

 11. $74.88 \neq 68.64$; false

Objective 3

 13. $x = 12$

 15. $x = 7$

 17. $x = 99$

 19. $x = 74$

 21. $x = 4$

 23. $x = 1.9$

 25. $x \approx 51.12$

 27. $x = 2\frac{1}{2}$

 29. $x = 7\frac{2}{9}$

6.4   Problem Solving with Proportions

Objective 1

 1. 15 ounces

 3. 4.69 meters

 5. 23.3 hours

 7. $295.98

 9. $23,945.64

 11. 45 pounds

 13. 165 people

 15. 17 milliliters

 17. 7000 students

6.5   Geometry: Lines and Angles

Key Terms

1. B	9. D	17. A
3. A	11. C	19. C
5. C	13. A	
7. C	15. B	

Objective 1

   1.(a) ray named $\overrightarrow{ST}$

   1.(b) line segment named $\overline{AB}$ or $\overline{BA}$

   1.(c) line named $\overleftrightarrow{LM}$ or $\overleftrightarrow{ML}$

Objective 2

   3. intersecting   **5.** parallel

Objective 3

   7. $\angle PRT$ or $\angle TRP$

Objective 4

   9.(a) obtuse

   9.(b) acute

   9.(c) straight; 180°

Objective 5

   11. intersecting

Objective 6

   13. $\angle AOR$ and $\angle ROM$;   **15.** $\angle WMT$ and $\angle TMV$;
       $\angle MOT$ and TOW $\angle TMV$ and $\angle VMN$;
                               $\angle VMN$ and $\angle NMW$;
                               $\angle NMW$ and $\angle WMT$

   17. 52°   **19.** 79° **21.** 168°   **23.** 20

Objective 7

   25. $\angle LOP$ $\angle MOQ$; $\angle POQ$ $\angle LOM$

Objective 8

  27. $\angle 1$, $\angle 4$, $\angle 5$, $\angle 8$ all measure 100°;
       $\angle 2$, $\angle 3$, $\angle 6$, $\angle 7$ all measure 80°

  29. $\angle 3$, $\angle 4$, $\angle 5$, $\angle 6$ all measure 125°;
       $\angle 1$, $\angle 2$, $\angle 7$, $\angle 8$ all measure 55°

6.6 Geometry Applications: Congruent and Similar Triangles

Key Terms

1. A
3. C

Objective 1

1. $m\angle 1 = m\angle 4$; $m\angle 2 = m\angle 5$, $m\angle 3 = m\angle 6$; $AB = DE, BC = EF, AC = DF$

Objective 2

3. ASA

5. SAS

7. SSS

9. SAS

Objective 3

11. $\dfrac{3}{5}; \dfrac{3}{5}; \dfrac{3}{5}$

Objective 4

13. $a = 20$ m; $b = 24$ m

15. $a = 10$ cm; $b = 2$ cm

17. 115.2 yd; 89.6 yd

19. 24 in.; 60 in.

Objective 5

21. 39 m

23. 25 ft

Answers to Worksheets for Classroom or Lab Practice

PERCENT

7.1   The Basics of Percent

Key Terms

1. C

Objective 1

1. 18%

Objective 2

3.(a) 0.15   **3.(c)** 2.60 or 2.6

3.(b) 0.08   **3.(d)** 0.003

Objective 3

5.(a) 70%   **5.(c)** 400%

5.(b) 13.5%   **5.(d)** 0.04

Objective 4

7. $\dfrac{3}{5}$ **11.** $\dfrac{1}{8}$ **15.** $\dfrac{1}{20}$

9. $\dfrac{3}{4}$ **13.** $2\dfrac{3}{4}$

Objective 5

17. 40%   **21.** 4%

19. 70%
                                         23. $11\dfrac{1}{9}\%$ or $\approx 11.1\%$

Objective 6

25. $63

27. 130 miles

29. $4\dfrac{1}{2}$ hours

7.2   The Percent Proportion

Key Terms

1. B

Objective 1

1. 12; $3000; unknown

3. unknown; 840 students; 336 students

Objective 2

5. $\dfrac{20}{100} = \dfrac{n}{4000}$; 800 voters

7. $\dfrac{p}{100} = \dfrac{19}{38}$; 50%

9. $\dfrac{42}{100} = \dfrac{357}{n}$; 850 books

11. $\dfrac{350}{100} = \dfrac{n}{10}$; 35 days

13. $\dfrac{75}{100} = \dfrac{n}{3260}$; 2445 athletes

15. $\dfrac{p}{100} = \dfrac{12}{40}$; 30%

17. $\dfrac{165}{100} = \dfrac{12.7}{n}$; 7.7 feet (rounded)

19. $\dfrac{p}{100} = \dfrac{207}{59}$; 350.8% (rounded)

21. $\dfrac{18}{100} = \dfrac{117}{n}$; 650 packages

23. $\dfrac{120}{100} = \dfrac{n}{80}$; 96 acres

25. $\dfrac{p}{100} = \dfrac{110}{78}$; 141.0% (rounded)

27. $\dfrac{140}{100} = \dfrac{1190}{n}$; 850 miles

7.3   The Percent Equation

Key Terms

1. D

Objective 1

1. $200 ÷ 4 = $50

3. 10 minutes ÷ 4 = 2.5 minutes

Objective 2

5. $21.04

7. 0.97 minute

9. $2.10

11. 0.097 minute

Objective 3

13. 833 letters

15. 950 students

17. 1230 acres

19. 42%

21. 400 magazines

23. 8%

25. 650 televisions

27. 9.5 liters

29. 600 cartons

7.4   Problem Solving with Percent

Key Terms

1. A

Objective 1

1. 652 students

3. 85%

5. 64 problems

7. 156%

9. 850 members

Objective 2

11. 125%

13. 20.1%

15. 5.4%

17. 53.5%

7.5   Consumer Applications: Sales Tax, Tips, Discounts, and Simple Interest

Key Terms

1. C
3. B
5. A

Objective 1

1. $8; $108

3. 2%; $99.92

5. $1.95; $34.44

7. $6\dfrac{1}{2}\%$; $15,123

Objective 2

9. $6; $6.47; $8; $8.63

11. $9; $9.58; $12; $12.77

13. $3; $3.59; $4; $4.79

Objective 3

15. $25; $75

17. $5.00; $44.99

19. 12%; $13.64

21. 15%; $27.20

Objective 4

23. $36; $336

25. $44.10; $1304.10

27. $4162.50; $22,662.50

29. $498; $8798

MEASUREMENT

8.1   Problem Solving with English Measurement

Key Terms

1. C

3. B

Objective 1

1.(a) 3   3.(a) 8

 (b)   60   (b)   12

 (c)   2   (c)   5280

Objective 2

  5. 6000 lb   7. 2 lb   9. 150 in.

Objective 3

  11. 2 pt
                      17. $3\frac{1}{2}$ gal or 3.5 gal   21. 432,000 sec
  13. 11,000 lb
                                                          23. $2\frac{1}{2}$ wk or 2.5 wk
  15. 2 hr   19. $\frac{3}{4}$ ft or 0.75 ft

Objective 4

  25. $2.59
                      27. $2\frac{3}{4}$ qt   29. $3.91

8.2   The Metric System—Length

Key Terms

1. D

3. B

Objective 1

1. m

3. km

5. mm

7. m

9. mm

Objective 2

11. 5000 mm

13. 8600 m

15. 0.050 m or 0.05 m

17. 830 mm

19. 0.009 m

Objective 3

21. 0.06 km

23. 5000 mm

25. 399 cm

27. 3.8 cm

29. 0.07 m

8.3   The Metric System—Capacity and Weight (Mass)

Key Terms

1. A

Objective 1

1. (a) L   3.   (a) mL

    (b) L   (b) L

Objective 2

5. 0.3 L   7. 37,150 mL   9. 16,000 mL   11. 76 L

Objective 3

13.   (a) g

       (b) mg

       (c) kg

Objective 4

15. 18,320 g   17. 300 mg   19. 9 kg   21. 93 mg

Objective 5

23. m   25. km   27. g   29. cm

24. mm   26. mL   28. L   30. kg

8.4   Problem Solving with Metric Measurement

Objective 1

1. $7.42

3. 4.03 m

5. 700 mg

7. 8.07 m

9. 6.25 L

11. 8 mm

13. $39.20

15. 70 mL

8.5   Metric–English Conversions and Temperature

Key Terms

1. B

Objective 1

1. 5.4 m

3. 3.2 gal

5. 32.7 yd

7. 7.1 lb

9. 127.6 g

11. 7.8 in.

Objective 2

13. 65°C

15. 14°C

17. 100°C

19. -8°C

Objective 3

21. −1°C

25. −4°C

23.   28°C

27.   54°F

29. −19°F

GRAPHS

9.1   Problem Solving with Tables and Pictographs

Key Terms

1. B

3. A

Objective 1

1. 57%

3. United

5. 3.1 luggage problems per 1000 passengers

7. United

9. $26.30

11. $16.33

Objective 2

13. 16 million people (or 16,000,000)

15. 7 million people (or 7,000,000)

17. 3 million people (or 3,000,000)

19. 40 million (or 40,000,000)

21. 25 million (or 25,000,000)

9.2   Reading and Constructing Circle Graphs

Key Terms

1. A

Objective 1

1. Age 2 years

3. 2 children

Objective 2

5. $\frac{1}{15}$

7. $\frac{6}{1}$

9. $960

11. home mortgage

13. $\frac{47}{100}$

15. 566 people

17. 1827 people

19. 3132 people

Objective 3

21. (a)   126°

(b).   54°

(c)   36°

(d)   90°

(e)   15%

(f)

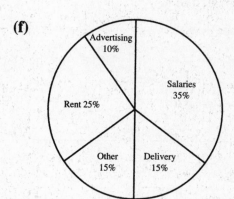

9.3 Bar Graphs and Line Graphs

Key Terms

1. A
3. D

Objective 1

1. 300 people

3. Watching TV; 900 people

5. 10%

Objective 2

7. May; 8500 workers

9. 1000 workers

11. 4500 workers; 112% increase

Objective 3

13. August; 600 tickets

15. 400 tickets

17. 250 tickets; 42% decrease

Objective 4

19. 35,000 computers

21. 10,000 computers

23. 25,000 computers; 63% increase

9.4   The Rectangular Coordinate System

1. C

3. A

5. D

7. C

9. E

Objective 1

1.–8.

9. (–1, 1)

11. (6, 4)

13. (–4, –4)

15. (2, 4)

Objective 2

17. II

19. no quadrant

21. IV

23. III

25. any negative number

27. any positive number

29. 0

9.5   Introduction to Graphing Linear Equations

Key Terms

1. A

Objective 1

1.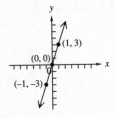

5.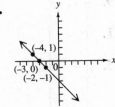

9.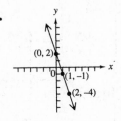

3.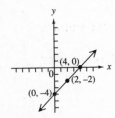

7.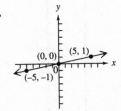

Objective 2

11. positive

13. negative

15. negative

9.5   Objectives 1 and 2

17.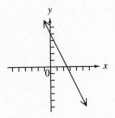

negative slope

21.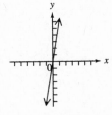

positive slope

19.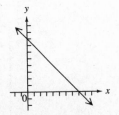

negative slope

23.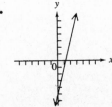

positive slope

EXPONENTS AND POLYNOMIALS

10.1   The Product Rule and Power Rules for Exponents

Objective 1

1. (a) x^5

   (b) $(-8)^3$

   (c) $(ab)^4$

3. (a)   64

   (b) -125

   (c) -81

Objective 2

5. 5^8

7. h^{14}

9. $-10r^7$

Objective 3

11. 7^{12}

13. t^{42}

15. $-v^{36}$

Objective 4

17. $256w^4$

19. $2x^7z^7$

21. $75m^2n^2$

Objective 5

23. $\dfrac{7^3}{3^3}$

25. $\dfrac{6^2}{5^2}$

27. $\dfrac{m^6}{g^6}$

29. $\dfrac{8^4}{y^4}$

10.2 Integer Exponents and the Quotient Rule

Objective 1

1. **(a)**   1

   (b)   −1

   (c)   1

Objective 2

3. $\dfrac{1}{6^2}$

5. $\dfrac{1}{x^7}$

7. $\dfrac{3}{10}$

Objective 3

9. 7^2

11. $\dfrac{1}{2^1}$ or $\dfrac{1}{2}$

13. 4^0 or 1

15. $\dfrac{1}{5^2}$

17. $\dfrac{1}{t^{12}}$

19. $\dfrac{1}{3^{20}}$

Objective 4

21. 9^2

23. $\dfrac{1}{b^1}$ or $\dfrac{1}{b}$

25. $\dfrac{1}{4^{10}}$

27. $\dfrac{1}{k^8}$

29. m^4

10.3 An Application of Exponents: Scientific Notation

Key Terms

1. A

Objective 1

1. 2.75×10^4

3. 3.86×10^7

5. 5.03×10^{-2}

7. 7.068×10^{-3}

Objective 2

9. 2,840,000

11. 0.09331

13. 760,000

15. 2005.2

Objective 3

17. 2.1×10^4

19. 4×10^{-6}

21. 8.4×10^{-19}

23. 2.125×10^4

25. 8×10^{-2}

Objective 4

27. 4.86×10^9 or 4,860,000,000 atoms

29. About 1.87×10^6 miles (rounded) or 1,870,000 miles per minute

10.4 Adding and Subtracting Polynomials

Key Terms

1. A   **2.** C   **3.** B   **4.** D   **5.** B   **7.** A

Objective 1

1. $-6y^9$ **3.** Already simplified **5.** $-5v^3 + 4v^2 + 8v$

Objective 2

7. $-4; 7; 7$

9. (a) trinomial

   (b) binomial

Objective 3

11. (a) -21 **13. (a)** -51

   (b) -26 **(b)** 74

Objective 4

15. $9b^3 + 2b^2$ **19.** $4x^2 + 6x - 18$

17. $-3c^2 + 6c + 2$ **21.** $3r^3 + 7r^2 - 5r - 2$

Objective 5

23. $8z^4 - 5z^2$ **27.** $-8w^3 + 21w^2 - 15$

25. $3s^2 + 7s - 3$ **29.** $5a^4 - a^3 - 6a^2 + 28a + 1$

10.5 Multiplying Polynomials: An Introduction

Objective 1

1. $8a^2 - 8a$

3. $18c^3 - 45c^2$

5. $-77n^5 - 44n^3$

7. $-30r^4 + 24r^3 - 54r^2$

9. $48s^5 - 28s^4 + 40s^3 + 8s^2$

11. $15k^6 - 50k^3$

Objective 2

13. $k^2 - 8k + 48$

15. $90b^2 + 47b - 77$

17. $12v^2 + 32v + 21$

19. $-84x^3 + 101x^2 - 61x + 90$

21. $-63u^4 + 106u^3 - 47u^2 - 10u + 8$

22. $12z^4 - 42z^3 - 89z^2 + 116z + 88$

23. $x^2 + 10x - 24$

25. $-7m^3 - 53m^2 + 28m + 32$

WHOLE NUMBERS REVIEW

R.1   Adding Whole Numbers

Key Terms

1. A

3. B

5. C

Objective 1

  1. 5   **3.** 17

Objective 2

  5. 18   **7.** 23

Objective 3

  9. 3970

Objective 4

  11. 135   **15.** 1171   **19.** 7126

  13. 124   **17.** 18,484   **21.** 7090

Objective 5

  23. 653 employees   **25.** 27 golf clubs

Objective 6

  27. correct

  29. incorrect; should be 9979

R.2   Subtracting Whole Numbers

Key Terms

1. A

3. C

Objective 1

   1. $8 - 6 = 2$ or $8 - 2 = 6$

   3. $6 = 5 + 1$

Objective 2

   5. minuend: 32; subtrahend: 23; difference: 9

Objective 3

   7. 224   **9.** 4212

Objective 4

   11. incorrect; should be 63   **13.** correct

Objective 5

   15. 11   **19.** 712   **23.** 47,639

   17. 433   **21.** 7809   **25.** 44,655

Objective 6

   27. $138

   29. 13,358 people

R.3   Multiplying Whole Numbers

Key Terms

1. A

3. product

5. C

Objective 1

  1. factors: 3, 5; product: 15

Objective 2

  3. (a)   120

     (b)   0

     (c)   18

     (d)   96

Objective 3

  5. 4896   **7.** 12,534   **9.** 155,259

Objective 4

  11. 13,080   **13.** 150,000   **15.** 210,000,000

Objective 5

  17. 1950   **21.** 89,397

  19. 20,160   **23.** 1,662,390

Objective 6

  25. 1888 hours   **27.** 140 words   **29.** 34,800 apples

R.4   Dividing Whole Numbers

Key Terms
1. B
3. C
5. B

Objective 1

1. (a) $\dfrac{52}{4} = 13$ and $4\overline{)52}$ with quotient 13

   (b) $38 \div 19 = 2$ and $\dfrac{38}{19} = 2$

Objective 2

3. (a) dividend: 8; divisor: 4; quotient: 2

   (b) dividend: 48; divisor: 16; quotient: 3

Objectives 3, 4, 5

5. 0   **7.** undefined   **9.** 1   **11.** 0

Objective 6

13. 42   **17.** 560 R3   **21.** 8735 R2

15. 1420 R5   **19.** 253 R2   **23.** 53,957

Objective 7

25. incorrect; should be 57,851 R2

27. correct

Objective 8

29. (a) 2, 5, 10

   (b) 3

R.5   Long Division

Objective 1

1. 214 R25   **5.** 298   **9.** 5214 R21

3. 62 R23   **7.** 5620 R7   **11.** 253 R556

Objective 2

13. 7   **17.** 40   **21** 6 R2700

15. 1800   **19.** 385

Objective 3

23. incorrect; should be 218 R16   **27.** incorrect; should be 1754 R5

25. correct   **29.** correct